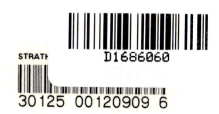
This book is to be returned on or before
the last date stamped below.

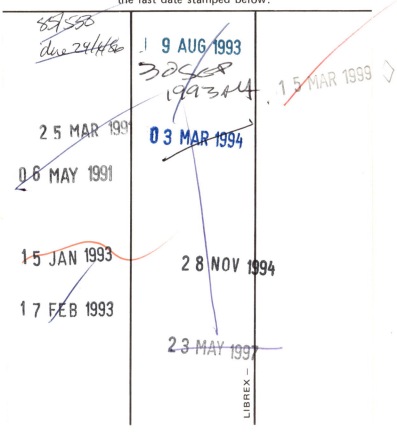

Energy Conservation in the Process Industries

ENERGY SCIENCE AND ENGINEERING:
RESOURCES, TECHNOLOGY, MANAGEMENT
An International Series

EDITOR

JESSE DENTON

Belton, Texas

LARRY L. ANDERSON and DAVID A. TILLMAN (eds.), Fuels from Waste, 1977

A. J. ELLIS and W. A. J. MAHON, Chemistry and Geothermal Systems, 1977

FRANCIS G. SHINSKEY, Energy Conservation through Control, 1978

N. BERKOWITZ, An Introduction to Coal Technology, 1979

JAN F. KREIDER, Medium and High Temperature Solar Processes, 1979

B. SØRENSEN, Renewable Energy, 1979

J. H. HARKER and J. R. BACKHURST, Fuel and Energy, 1981

STEPHEN J. FONASH, Solar Cell Device Physics, 1981

MALCOLM A. GRANT, IAN G. DONALDSON, and PAUL F. BIXLEY, Geothermal Reservoir Engineering, 1982

W. F. KENNEY, Energy Conservation in the Process Industries, 1984

In preparation:

CHUN H. CHO, Computer-Based Energy Management Systems: Technology and Applications, 1984

Energy Conservation
in the Process Industries

W. F. Kenney

Chemical Technology Department
Exxon Chemical Company
Florham Park, New Jersey

1984

ACADEMIC PRESS, INC.

(Harcourt Brace Jovanovich, Publishers)
Orlando   San Diego   San Francisco   New York   London
Toronto   Montreal   Sydney   Tokyo   São Paulo

ACADEMIC PRESS, INC.
Orlando, Florida 32887

United Kingdom Edition published by
ACADEMIC PRESS, INC. (LONDON) LTD.
24/28 Oval Road, London NW1 7DX

Library of Congress Cataloging in Publication Data

Kenney, W. F.
    Energy conservation in the process industries.

    (Energy science and engineering)
    Includes index.
    1. Industry--Energy conservation. I. Title.
II. Series.
TJ163.3.K46 1983   621.042   83-11937
ISBN 0-12-404220-1

PRINTED IN THE UNITED STATES OF AMERICA

84 85 86 87   9 8 7 6 5 4 3 2 1

Contents

Preface

Even in periods when oil and energy in general are not escalating in price, or when the world may be in the grip of a recession which markedly reduces energy demand, and even with appreciable real progress toward more efficient use of energy in all facets of the economy already accomplished, a book on industrial energy conservation is not superfluous. In spite of all the reduction in energy demand, the industrial nations are still importing large quantities of oil each day. In addition, the hiatus in oil price increases has caused the deferral of many projects aimed at switching to coal, shale, and other indigenous energy souces perhaps by as much as a decade. Thus I have concluded that energy supply is still a problem for the industrial nations (and the world). And, if supply is a problem, can economics be far behind?

As for energy conservation progress thus far, I believe only the surface has been scratched in the process industries. At present, inefficient units have been shut down. The conservation measures implemented on those units still operating are largely in the realm of housekeeping and simple additional heat recovery. Few fundamental improvements in process or equipment design have been implemented. These will be necessary for dealing with the economics of shortage in the future.

Thus the thrust of this book is to provide some insight into ways of identifying more significant energy efficiency improvements. This involves a marriage of practical and fundamental principles, of academic and industrial scenes, of entropy and economics. I believe that this can be done, however imperfectly at the start.

Although the book deals mainly with principles, it is meant to be practical. It draws from material used in continuing education courses on industrial energy conservation and integrates a great deal of practical experience to present a consolidated picture. Many examples are given, but it is *not* a comprehensive handbook from which specific application can be drawn in all plants. The purpose of the examples is to demonstrate how the principles can be used to practical advantage. They are also meant to encourage readers to apply the fundamentals to their own plant situations.

The second law of thermodynamics wears many masks in the industrial environment, yet much of what is done in the name of energy conservation can be related to its principles. Even so, this is not a text on thermodynamics or on the methodology of second law calculations. Rather, the book is about some ways the teachings of the second law are and can be further applied for economic benefits in the process industries. The insights are limited to what I could reconcile with my own experience. The level of mathematical sophistication is also constrained to what I have found to be easily manageable in an industrial setting.

After giving an overview of the current energy situation and a brief refresher in thermodynamics, the book describes a staged approach to improved energy use, each stage more complex than the last. The first stage (Chapters 3–5) is to understand where the energy goes and how to calculate the value of losses. It also provides some insights on how to get the most out of existing facilities. This stage might be called "establishing the optimum base case." The second stage (Chapter 6) involves improving facilities based on an understanding of the overall site energy system. Again, principles are stressed. The third stage (Chapters 7–10) involves the identification of fundamental process and equipment improvements. Here, making and using second law analyses are explicitly discussed, along with some preliminary observations about the relationships between efficiency and capital cost. The book closes with chapters on more systematic and sophisticated design methods and some guidelines and checklists that can be distilled from the text. The former relate thermodynamics and economics to integrate the two essential value systems for the process designer–improver. The objective is to give more guidance to creative thinking than is possible from the thermodynamic viewpoint alone. Chapter 10 is both a summary and a reaffirmation that even sloppily directed action is more likely to generate profits than doing nothing.

In writing this book I have brought together my interpretation of many studies, some of which were done for Exxon. In addition to the references in the text, the list of those who contributed one insight or another is too long to present here, and the chance for omitting a worthy name is too large for me to risk. In addition, I had the privilege of being taught by some of the most prestigious chemical engineering thermodynamicists of the 1950s and of having recent discussions with a number of others. Little did any of us know it would come to this. I am grateful to all who helped me and especially to Exxon Chemical for permission to publish this work.

List of Common Symbols

Variables

A	available energy (sometimes equivalent to exergy, essergy, availability, etc.)	Q	heat flow
		R	gas constant = 1.987 Btu/mol°R, the ratio (power requirement)/(process heat from steam)
C	heat capacity, capital cost, flow rate of chemicals	S	entropy
E	internal energy	T	temperature
F	fuel flow rate	U	overall heat transfer coefficient
G	Gibbs free energy	V	volume, thermoeconomic flux
H	enthalpy	W	work
KE	kinetic energy	X	mole fraction
L	cost of labor	α	price of a parcel of available energy
M	mass flow rate	γ	fugacity (activity coefficient)
N	number of moles	Δ	difference between inlet and outlet properties
P	pressure		
PE	potential energy	η	fuel utilization or efficiency

Subscripts and Superscripts

(others may be defined in the text)

A	combustion air	LP	low-pressure steam
act	actual	min	minimum
atm	atmosphere	mix	mixing
B	boiler	O	ambient (outside)
bot	bottoms	p	power
C	condenser	ph	preheated air
cs	closed system	proc	process
CW	cooling water	prod	produced
dest	destruction	R	reaction
env	environment	r	reboiler
F	fuel, furnace, flame	ref	refrigerant
f	of formation, final	rev	reversible
FG	flue gas	S	stack
H	heat	stm	steam
HP	high-pressure steam	sur	surroundings
i	ideal	thr	throttle
irr	irreversible	tot	total
L	leaks, latent	v	vapor
liq	liquid	w	makeup water
LM	log mean		

1

Energy Outlook

Introduction

The fundamentals of energy conservation include stoichiometry, thermo-dynamics, and economics. The term "thermoeconomics" is not used in this context because, as we shall see later, it is generally reserved for use in a more specific sense. This book should not be misconstrued as a text in any of the underlying fundamentals. The book does, however, deal with the application of these principles to industrial problems in energy conservation. Its primary objective is preparing engineers in industrial process research and development, design, manufacturing, and energy conservation to combine these fundamentals in a practical way.

Emphasis will be on the application of the science to practical problems. The industrial environment involves constraints that are other than scientific or economic. More will be said about these later, but they are every bit as important as the scientific constraints in achieving practical solutions.

The fundamentals discussed apply to all industries. Examples in the text are drawn primarily from the chemical and refining industries and their associated utility systems. As such, the primary thrust is toward continuous processes, but some comments on batch operations are also included.

I. Scope of the Problem

A. The Impact of the Oil Embargo

The oil embargo of 1973 prompted raw material and energy conservation activities by industrial concerns throughout the Western world. Many companies had already initiated some conservation programs, but the rapid escalation of oil prices drew everyone into action. In addition, the availability of supposedly secure supplies of oil and natural gas also came into question at a time when the use of coal was being restricted because of environmental considerations. In short, dramatic events combined to create a near panic environment in the industrial sector of the economy as well as in the lines at local gas stations.

Through the Chemical Manufacturers Association (CMA), the American Petroleum Institute, and other trade organizations, the government extracted energy conservation promises from various segments of industry. These voluntary targets typically projected a 15% saving in energy use per unit of production for 1980 when compared with the base year 1972. Companies immediately launched housekeeping and rehabilitation activities (such as repairing steam traps and insulation), started motivation programs for their wage earners, and installed simple heat exchange systems to recover wasted energy at all levels. Since much of our industrial base was designed when fuel was very cheap, it is not surprising that these efforts, spurred by rapidly escalating costs, were more than enough to reach the government-sponsored targets ahead of schedule. The current CMA conservation target for 1985 is 30%. Data for the end of 1981 already showed 24% less energy consumption per unit of production, indicating the new target would likely be reached as well.

In Japan even more rapid improvements were made. The government imposed arbitrary cuts in absolute energy use during the embargo. Companies could produce as much as they liked as long as their total energy consumption was 15% lower than their pre-embargo level. In spite of a long tradition of frugal energy use by Japanese industry, many plants in Japan were able to meet the government requirements with little or no cut in production.

B. Costs Outstrip Conservation

While a moral victory was being achieved, the economic war was being lost. Figure 1-1 shows the energy cost profile from 1972 to 1980 of a typical U.S. Gulf coast petrochemical plant. The data are based on a 1977 forecast of energy costs and of the effect of energy conservation programs. Feedstock costs are excluded from this figure. Energy costs in 1972 were $16 million (18% of total operating costs). Energy costs in 1980 were more than 10 times this figure and 58% of the total, even though energy savings equal to the entire 1972 operating cost were achieved by energy conservation programs. Two other examples, shown in Table 1-1, demonstrate similar economic impacts in spite of energy savings of 24 and 37%/unit of production.

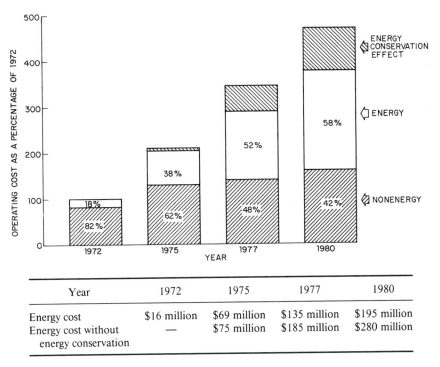

Year	1972	1975	1977	1980
Energy cost	$16 million	$69 million	$135 million	$195 million
Energy cost without energy conservation	—	$75 million	$185 million	$280 million

Fig. 1-1 A comparison of energy-related and non-energy-related operating costs (excluding feedstock costs) for a typical U.S. chemical plant, including the effect of energy conservation.

The net result of the early years of energy conservation activity was to reduce by a relatively small amount (about 25%) the ravages of rapidly escalating energy costs on the profitability of processing plants.

The conclusions from these data are as follows.

(1) The energy conservation targets initially advanced by the trade groups were woefully inadequate.

(2) A much more fundamental approach to energy conservation is required if the economic impact of increasing energy costs is to be dealt with effectively.

C. Continued Cost Pressure Is Likely

At this writing there is a lull in the rate of escalation of crude oil prices. Economists attribute this to the elasticity of oil demand as a function of price and to the economic downturn. No one can predict how much of this

Table 1-1

Examples of Energy-Cost Variation over Time[a]

Industry	Energy Cost				
	Actual			Projected	
	1972	1979	1980	1985	1990
Manufacturing concern	—	977	1135[b]	2200	—
Petrochemical company	100	—	500	1500[c]	3000

[a] In millions of dollars.
[b] After a 23.5% savings per unit of production.
[c] After a 37% savings per unit of production.

change in price is permanent (elasticity) and how much cost pressure will be produced when business turns back up. Indeed, petroleum demand has decreased much more than originally anticipated when its price rose. Because of economic need, all sectors of the economy have reduced consumption by more than the original token amounts offered to the government. Such activities were assisted by a steeper increase in oil prices than in capital costs over part of that period.

In Fig. 1-2 an attempt is made to show a relationship between the escalation of crude oil prices and the inflation of capital costs since 1977. The slopes of the two curves are parallel for the first 2 yr of the period, but they diverge thereafter in a new round of oil price increases. Justifying capital investment to save energy was relatively easy under such circumstances. This cannot continue, however. Indirectly, energy cost increases

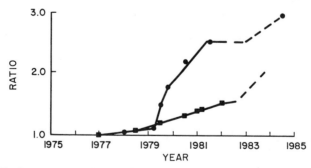

Fig. 1-2   Equipment costs and crude oil prices, a comparison of inflation rates: ●, the ratio of crude oil prices over time to the price in 1977; ■, the ratio of the Marshall Stevens Equipment Cost Index over time to the cost in 1977; ---, projected values.

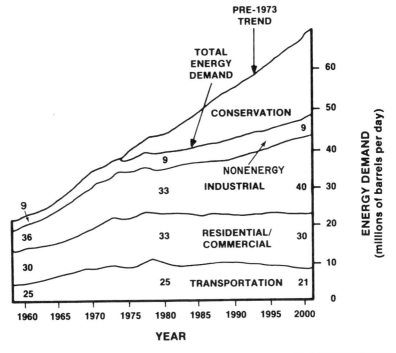

Fig. 1-3 A 1981 prediction of U.S. energy demand by consuming sector.

Sector	Growth rate (%/year)			
	1960–1973	1973–1980	1980–1990	1990–2000
Nonenergy	4.1	—	1.3	1.4
Industrial	3.7	−0.1	1.1	2.6
Residential–commercial	4.6	0.9	0.7	0.1
Transportation	4.1	0.3	−0.3	0.3
Weighted average	4.1	0.3	0.7	1.2

will increase the cost of capital goods through the manufacturing cost segment. Workers are clamoring for catch-up pay increases to offset personal energy costs. The potential exists for capital costs to resume a rate of increase more nearly parallel with that of crude oil prices. If this occurs, reducing energy consumption through incremental capital investment will be that much more difficult.

Supply–demand relationships for petroleum also presage continued cost pressure over the long run. Figure 1-3 shows the U.S. energy demand

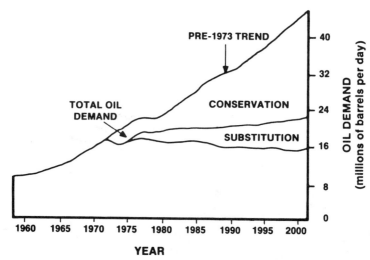

Fig. 1-4 A projection of U.S. oil demand to the year 2000.

situation projected by Exxon Corporation in 1981.[1] Total energy demand is projected to increase on through the year 2000. Industrial energy demand as a share of the total will also increase (to 40% vs. 30% in 1980), whereas other sectors will decrease their share of the total demand. This situation is projected in spite of the energy conservation activities of the industrial sector.

U.S. *petroleum demand* is projected to decrease slightly (9%) at the turn of the century relative to 1980 (see Fig. 1-4). Most of the conservation effort is seen in the reduced oil demand, and an appreciable substitution of other fuels for oil is anticipated. However, 16 million barrels/day (B/d) of petroleum are still likely to be needed in the United States by the year 2000. It is anticipated that much of the substitution for oil will occur in the electric utility, industrial, and commercial sectors of the economy. The transportation and nonenergy segments of the economy have very limited potential for reducing petroleum consumption.

Even the United States, which is better off in petroleum supplies than many other countries, can supply only about half its projected demand domestically (see Fig. 1-5). Production from known reserves has long since peaked and is currently running at about half of its 1970–1975 peak level. Appreciable production from new discoveries is anticipated in the 1990s. A significant contribution from synthetic petroleum was expected in the same time frame, but this will probably be delayed for a decade by recent decisions to cancel or defer major investments. The result is that in the 1990s

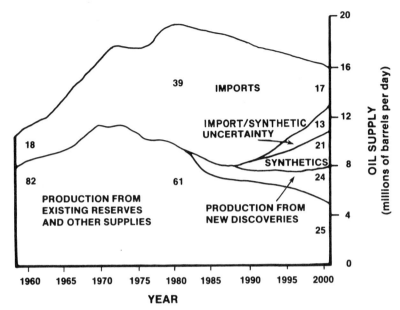

Fig. 1-5 A projection of the U.S. oil supply by source to the year 2000.

imports and additional conservation on the order of 8 to 10 million B/d will be the only way to close the supply–demand gap.

Recent data and projections of U.S. oil demand are somewhat lower than those shown in Fig. 1-5. For example the U.S. Energy Information Administration has projected the 1983 average oil demand at 15.4 million B/d, up 0.7% from 1982, based on a 1.7% increase in Gross National Product (GNP). This increase in demand is projected to cause an 11.6% increase in oil imports to 4.7 million B/d, indicating the inflexibility of the U.S. oil supply. Current major oil company capital expenditures are down 25% from 1982, which is impacting exploration budgets. The 1984 U.S. demand outlook is 16 million B/d, up 5% over 1983 and approaching the amounts shown in Fig. 1-5. Barring a major technology breakthrough, the United States is likely to continue consuming more petroleum than it can produce and thus remain heavily dependent on imports to cover changes in demand.

Special mention is in order for the petrochemical business. The 1970s saw markedly reduced growth in the petrochemical industry because of the rapid increase in feedstock costs. Projected growth in the 1980s, however, may still be higher than the growth rate forecast for the gross national product.[2] In 1980 the industry produced 80 million tons of product, which generated worldwide revenues of over $200 billion. In the United States petrochemical feedstock represents only 3% of petroleum demand. But with decreasing petroleum supplies, significant growth in the petrochemical business will require very careful feedstock selection and management in future years. Thus, we are seeing major ethylene producers rapidly turning to gas liquids for feedstock. The question is, "Will there be adequate supplies once the current recession is over?"

From these arguments it seems reasonable to conclude that, apart from temporary lulls in price escalation, petroleum costs will continue to increase over the long run. This view is shared by a number of investors in the United States who are buying unprecedented amounts of oil-income funds. Through these funds groups of investors underwrite production costs from known reserves in return for a share of the profits: they are betting their money on a long-term upward trend in prices.

D. Price Projections

Long-lived petroleum price forecasts are impossible to make. Psychological and political activities play a major role in price forecasts, and corporate planning departments tend to overreact to these signals because of a history of being burned. The author can do little but sympathize with the plight of those who undertake to project energy costs and offer the caveat that the following forecasts were wrong the day this book went to press. However, some general feel for likely future prices was thought to be of value in providing perspective if nothing else. Table 1-2 gives two pricing projections for Arab Light marker crude oil (FOB Saudi Arabia) in dollars per standard barrel through the rest of the century. Both the price in the year projected (i.e., current dollars) and the equivalent 1980 prices are shown in the 1980 outlook. The average escalation rate for the last two decades of the century is about 3%/yr in 1980 dollars. An economic advisory service has predicted a 2–4%/yr increase over the same period following a lull in the first half of 1982, and Chase-Manhattan Bank projected an average increase of about 2.8% based on a 0.7% increase until 1985 and a 4.5% increase for the remainder of the decade.[3]

The 1982 projection in Table 1-2 is significantly lower in current dollars. A longer plateau in prices is anticipated, along with a lower escalation rate. The average price of crude oil in the United States will probably be about 10% higher than the price of Arab Light but will depend on the mix of domestic production. European and U.S. prices for most products are not projected to be much different, but Canadian prices will most likely be maintained lower by government policy.

The 1982 price forecast was considered by the author to be an "acid test"

Table 1-2

A Comparison of Pricing Projections for Arab Light Marker Crude Oil[a]

Year	1980 Outlook		1982 Outlook (current dollars)
	Current dollars	1980 dollars	
Actual			
1977	12.40		
1978	12.70		
1979			
First quarter	13.34		
Second quarter	15.70		
Third quarter	18.00		
Fourth quarter	22.00		
Projected			
1980	27.5		
1981	31	28	
1985	56	37	37
1990	90	40	68
2000	199	45	160

[a] In dollars per standard barrel.

for energy conservation investments for most of 1982. Recently even lower crude-oil price forecasts were advanced by the president of Shell Oil.[4] He spoke of price increases to $51 per standard barrel in 1990 (as compared with $68 in the 1982 forecast) and increasing imports. This may represent a low-side sensitivity for planning purposes.

On the other hand, current turbulence in the Middle East has prompted considerable fear of rapid price escalation. If Iran were to block the Strait of Hormuz, the price of oil might reach $65–130/B very quickly, even though only 4.5% of U.S. oil passes through the strait. Such is the force of emotional (rather than economic) pressure on oil prices. The question is how long the panic might last.

The overall trend in current price projections for petroleum is to lower future values in spite of a general acknowledgment that a higher percentage of future supplies must come from OPEC. The author feels that the underlying supply–demand structure will force prices higher rather than lower in the long run, but the timing is difficult to establish. Support for this point of view comes from Richard Balzhiser,[5] who supports projections based on a 2.5%–3.0% per year real oil price increase through the rest of the century.

Approaching crude prices from the point of view of how much greater the rate of price increase will be than the rate of inflation has a number of advantages. Oil price projections can be reworked very simply as estimates of inflation rates change. In addition, best- and worst-case scenarios can be constructed simply by using different assumptions for the real price increase.

Another of the great uncertainties at the moment is how much of the current decrease in oil demand is due to real conservation and how much to business recession. Obviously the latter will disappear when business improves and may fuel another upturn in prices, whereas the former represents a real gain in process efficiency. Since companies tend to shut down their least efficient units first when existing capacity is not needed, a business upturn might cause a more than linear increase in unit energy demand. The answer to this riddle will not be *known* beforehand, in spite of sophisticated calculations. This industrial energy conservation engineer needs to develop a list of improvements that is well understood and flexible enough to allow continued implementation in an environment with fluctuating price outlooks.

It is generally accepted that the total economic cost of importing large volumes of oil is much higher than the price of the oil. The resulting trade deficits fuel inflationary pressures and influence the ability to pay for other imports. As the OPEC countries move downstream into refining and petrochemical manufacturing, products, exports, and jobs are being redistributed away from the industrialized countries to the oil-rich countries. The limits of this trend are nowhere in sight at this time.

E. Coal Is Characterized By High Investment Costs and Low Energy Costs

On an energy basis, coal prices tend to run at about 40% of the price of low-sulfur fuel oil. Environmental and handling problems make investment costs for coal-based energy systems much higher than those for systems fueled by petroleum and gas. Large, new deposits in addition to known reserves in North and South America indicate that many years' supply of coal exists. However, the technology of coal utilization is, in general, old. Considerable room for improvement exists and much research and development is currently underway.

As indicated previously, the use of coal generally means very large fixed costs but much lower incremental energy costs. For example, a major coal-fired cogeneration project costs several billion dollars, but *incremental* steam costs are still lower than those for an oil-fired boiler. Thus, coal-fired systems will likely approach the kind of economics prevalent in the nuclear power industry, in which base-loading the plant is a primary objective. In

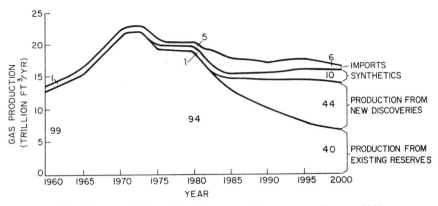

Fig. 1-6   A projection of the U.S. gas supply by source to the year 2000.

addition, the drive to convert coal to liquid and gaseous fuels to protect previous investments in processing facilities designed for these fuels will be strong, even though the thermodynamics of this processing step involves appreciable energy losses. Here, the relationship between thermodynamics and economics is emphasized. Thermodynamically less efficient systems may well be economic because of the large fixed costs associated with installing a major coal supply and handling facility and dealing with the environmental impact of coal utilization.

F. Gas Supply – Demand Projections

Gas supplies tend to be more restricted to local use than either oil or coal supplies. The cost of the long pipelines for natural gas distribution tend to be much larger than those for oil. Moreover, many of the new natural gas finds have occurred during the period since the embargo. As a result, governments tend to control the distribution of this valuable natural resource much more stringently than they did a generation ago. Figure 1-6 shows a projection of the U.S. gas supply situation. Natural gas demand has declined in the last two decades, but the industrial share is projected to increase to about 55% of the total by the year 2000. Projections of gas reserves tend to vary somewhat more than those for petroleum reserves owing to the varying judgments of the practicality of recovering gas from very deep wells, of the economics of gas gathering systems, etc. More than half of the domestic gas available in the year 2000 is projected to come from future discoveries. Production from existing reserves has been falling off rapidly since 1981. Synthetic gas production will not become significant until the end of the century, and

Table 1-3

Actual and Projected U.S. Natural Gas Prices[a]

Source	Actual			Forecast			
	1978	1979	1980	1984	1985	1986	1990
Intrastate (wellhead)	179	205	237	436	762	879	1365
Interstate[b] (wellhead)	96	120	164	336	506	618	1083
Imported (border)	224	250	403	685	762	879	1365
Interstate (city gate)	146	174	222	426	605	733	1260

[a] In current cents per million Btu. From Reference 6.
[b] Includes imported natural gas.

perhaps later if current outlooks are correct. Imports will represent approximately the same percentage in the year 2000 as in 1980, with some increase in the intervening 20 yr.

Gas prices can be expected to vary widely from region to region. Regions such as Saudi Arabia, the countries around the North Sea, Alberta, Canada, and Indonesia all seem to be adopting different policies with respect to the pricing of gas. These gas-rich areas are all attracting consideration from petrochemical producers, who are interested in converting the ethane and propane constituents of the gas into various plastics and other chemicals. In some cases fuel and feed will be priced very low to attract investment into riskier areas, whereas in others prices may well be comparable to those in areas where gas supplies are much less available. Volatility would seem to be one common characteristic of gas price projections.

With this caveat in mind, a recent U.S. natural gas price projection[6] is given in Table 1-3. Decontrol of prices in 1985 is assumed. "City gate" prices include an estimate of pipeline transportation cost added to the well-head price for interstate gas. Domestic gas prices are expected to increase sixfold from 1980 to 1990 in 1979 dollars. Most of this increase is assumed to occur in the latter half of the decade.

For those interested in energy efficiency, an additional challenge is presented by the variability of gas pricing and its impact on petrochemical feedstock prices around the world. In low-cost areas little capital can be justified for use in improving feed and fuel utilization. Yet, the value of the material to the rest of the world remains high. Thus, the challenge becomes to maximize efficiency while minimizing the capital required to achieve it. The pressure of these situations may well bring about the development of better technology than is presently available. It is in these areas, in particular, that a sound, fundamental analysis of the situation shows promise for maximum benefits.

Table 1-4

Comparison of the Specific Fuel Consumption of Known Processes with the
Theoretical Minimum for Selected U.S. Industries[a]

Process	Specific fuel consumption 1968 (Btu/ton)	Potential specific fuel consumption using 1973 technology (Btu/ton)	Theoretical minimum specific fuel consumption[b] (Btu/ton)
Iron and steel manufacturing	26.5×10^6	17.2×10^6	6.0×10^6
Petroleum refining	4.4×10^6	3.3×10^6	0.4×10^6
Paper manufacturing	39.0×10^{6c}	23.8×10^6	$> -0.2 \times 10^{6d}$ $< 0.1 \times 10^6$
Primary aluminum production[e]	190×10^6	152×10^6	25.2×10^6
Cement production	7.9×10^6	4.7×10^6	0.8×10^6

[a] From Reference 7.
[b] Based on thermodynamic availability analysis.
[c] Includes 14.5×10^6 Btu/ton of paper produced from waste products consumed as fuel by the paper industry.
[d] A negative value means that no fuel is required.
[e] Does not include the effect of scrap recycling.

II. Thermodynamic Efficiencies

A. Industrial Efficiencies Are Generally Low

Much more will be said later in this book about the definition of thermo-dynamic efficiencies. At this point only a word of introduction will be given.

The concept of thermodynamic efficiency and the so-called "second law analysis" is based on the concept of available energy as defined by Gibbs and developed by Keenan and others. The available energy function is a state variable of a material. In the simplest case it is defined as follows:

$$A = H - TS,$$

where A is the available energy function of the material at the given conditions, H the enthalpy, T the temperature of the system, and S the entropy of the material. These definitions exclude terms that would cover the concentration differences between the material and the environment, and terms describing potential and kinetic energy effects. These terms will be included in the definition presented in Chapter 2.

Available energy has been called by many names in recent energy efficiency studies. These include "availability," "capacity to cause change," and hosts of related terms, such as "essergy." The available energy of a material at a given set of conditions represents its capacity to do work or cause change relative to the environment. It measures the *quality* of the energy contained in the material.

Gyftopoulos, Lazaridis, and Widmer, in their study "Potential Fuel Effectiveness in Industry," compare 1968 technology for various industries with more modern technology, with the theoretical minimum specific fuel consumption for the process calculated by availability analysis.[7] This comparison is presented in Table 1-4. To obtain the fourth column in this table, the minimum thermodynamic availability change required to produce the products in question was calculated.

Two conclusions are obvious from the data in this table. The first is that, for the industries in question, appreciable immediate energy efficiency improvement is possible by the application of existing technology. The second conclusion is that even the improved operation is still a long way from the best available energy utilization that is theoretically possible.

Of course the theoretical efficiency is not achievable in practice, because it is based on infinitesimal driving forces and machinery that does not exist. Indeed, information obtained from theoretical analyses may often appear to be esoteric to the industry's technical people and downright useless to management. However, the large differences between the efficiency of pre-embargo technology and the theoretical thermodynamic efficiency indicate a very large potential for process improvement. The challenge is to utilize thermodynamic analysis in a practical way and, equally important, to communicate the results in a way that will convince management to act rather than scoff. This is entirely possible through simplified approaches and careful presentation of results.

An interesting practical example of a giant step forward is the Unipol process licensed by Union Carbide. Here, the production of low-density polyethylene plastics is accomplished by a low-pressure process rather than the conventional 50,000-psig route. In this process both the energy efficiency and the capital costs of a plant were improved remarkably by the development of a new catalyst system. Similar improvements may be possible in other industrial processes as well.

Various energy policy studies performed in the mid-1970s were the first to develop data showing that on a global scale the required investment for efficient facilities is less than that for a number of conventional processes and their associated energy systems. Chapter 10 is devoted to the capital cost aspects of energy efficiency.

B. Science in the Industrial Environment

In the United States the obvious constraint on the industrial application of science is that it must be profitable. However, there are many other constraints, both external and internal, that can prevent achieving the best potential energy efficiency in an industrial plant, even if economic criteria are satisfied.

External constraints are often imposed by governments, the requirements of the Environmental Protection Agency being perhaps the most obvious. But community relations, business cycles, relations with utility companies and industrial or commercial neighbors, geography, climate, and institutional and governmental politics are all factors that must be considered. Under normal circumstances many of these factors can be handled with the application of the proper talent, but ignoring these problems is a sure route to failure.

Most internal constraints have to do with people: either enough talent is not available or new ideas cannot be sold to decision makers. If the key leadership role can be filled within a company, shortages of other talent can be made up through consultants and/or simply allowing more time in the schedule. Remember, given access to the right skills, technical efforts can often be stretched out without loss of efficiency and with the side benefit of greater opportunity for creative thinking.

Regardless of technical merit, it will almost always be *necessary to sell your ideas* for improving energy efficiency. Basically, most companies in the process industries are not in the energy business; their prime objectives are to produce products, and energy is only one of several cost factors. It is not unusual for company policies to require higher hurdle rates for energy projects. It is not unusual for plants to pursue minor improvements in capacity or product quality while allowing more profitable energy improvements to languish for lack of resources. It is not unusual to encounter skepticism and doubt about new technology, or even an active effort to discredit the technology proposed. The only path around these obstacles is to recognize that these facets of the problem are as valid as the technical and economic ones. Work as intelligently and persistently on them as you did on the other aspects, and progress will result. Strategies that have been found effective and that have a sound basis in management science include

(1) finding a qualified champion,
(2) working with the end user to develop support,
(3) establishing a good track record in smaller issues,
(4) capitalizing on interactions with other programs that are viewed positively,
(5) getting the proper visibility.

You should also recognize that compromise may well be the right answer, particularly for a first application.

You will know that progress is being made on other issues when the crusty old operations manager reaches for his big gun: *Safety.* Often improved energy efficiency means greater complexity or less margin for error for the operators. This is rarely attractive to those responsible for operations, particularly if flammable or toxic materials are being handled. In addition, an enthusiastic energy conserver may overlook real safety problems in his zeal to sell his ideas. *A thorough safety analysis of the proposed technology is essential.* This provides an opportunity to implement some of the strategies mentioned above. Make the safety analysis a joint effort with the operations department to gain their commitment and to ensure yourself that the proposal is safe, even under abnormal or transient conditions. If necessary, plan and carry out appropriate demonstrations of operability and safety aspects. There is no substitute for understanding any new or worsened risks and providing effective countermeasures.

On the other side of the ledger, energy efficiency and environmental issues often blend synergistically. Reducing the amount of fuel fired will reduce SO_2 and NO_x emissions, boiler and cooling tower blow-down, and thermal discharge. These interactions can provide additional credits and broader organizational support for energy conservation projects. Be alert for such opportunities, and be sure that they are communicated effectively to management.

From time to time in the text we shall return to this theme of effective communication to management. Many of the frustrations mentioned previously can be minimized if management understands the importance of energy efficiency to the "bottom line." Although this book concentrates on the technical aspects of energy efficiency improvement, little progress will be made without management commitment.

III.   The Fundamental Strategy

The hiatus in fuel price increases provides the time to develop and install various kinds of fundamental energy efficiency and process improvements. In good times this will provide an opportunity for increased profits; in bad times, the competitive edge needed to survive. A practical approach to energy conservation in the industrial environment can be split into four components. These are given in the following list. Each component is not carried out in series, but rather considerable overlap is required both to keep up with cost pressures and to ensure the optimum mix of the efficiency-improvements.

(1) The first step is to know how your energy is used and what it is really worth to you. The worth of each block of energy may well be different from the price you currently place on it. A fundamental understanding of the available options is the only way to establish the true worth of any parcel of energy.

(2) The second strategy is to get the most of your existing process and energy system. This includes better control arrangements and facilities maintenance and the ability to capitalize on variations in throughput, weather conditions, etc. A full understanding of process interactions, as well as the optimum level of manpower and maintenance activity, is essential to achieve this goal.

(3) The third strategy is to optimize the plant process and energy system. To the extent possible, beneficial interactions with industrial or commercial neighbors should be brought into the system to maximize the potential for energy efficiency through cogeneration. In this step all of the energy suppliers and users are drawn together in a single optimized network. Political, institutional, and operational constraints must be woven into the thermodynamic and economic analysis to provide a viable energy plan for the site and those neighbors that are included. This plan must look ahead at least 10 yr to account for longer-range changes in fuel costs and/or type so as to remain competitive as external constraints change.

(4) While all of this is going on, efforts must be made to develop fundamentally improved processes for the company's "bread and butter" products. Fundamental energy analysis of the existing processes can assist in directing research and defining the potential for improvement. This will help focus the creative thinking of R&D personnel but will not substitute for it.

Notes to Chapter 1

1. R. B. Nesbit, "Energy and Petrochemicals," p. 1ff. Exxon Chemical Americas, Houston, Texas, 1981.

2. S. Field, Ethylene prices in the United States. *Chemical Engineering Progress* **69,** 22 (1982).

3. Slower oil price, spending rise seen. *Oil and Gas Journal* **80,** 50 (1982).

4. Shell: weak crude oil prices could reduce list of projects. *Oil and Gas Journal* **80,** 36 (1982).

5. R. B. Balzhiser, U.S. energy prospects. *Chemical Engineering* **90,** 72 (1982).

6. S. Field, Ethylene prices in the United States. *Chemical Engineering Progress* **69,** 19 (1982).

7. E. Gyftopoulos, L. J. Lazaridis, and T. F. Widmer, "Potential Fuel Effectiveness in Industry." Ballinger, Cambridge, Massachusetts, 1974.

2

The Second Law of Thermodynamics Revisited

Differences between Laws

To understand the fundamentals of the efficient use of fuel, we must consider the quality as well as the quantity of the energy used to achieve a given objective. Table 2-1 shows two different views of energy efficiency for a number of processes and unit operations.[1] These efficiencies are loosely labeled "first law" and "second law," referring to the corresponding laws of thermodynamics.

To refresh the reader's memory, the first law of thermodynamics states that energy is conserved. More specifically, it states that the enthalpy contained in all of the input streams to a process must appear somewhere in the output streams from the same process. One such output stream could, of course, be a loss to the atmosphere or other heat sink. Thus, the energy efficiencies listed in the "first law" column of Table 2-1 represent the enthalpy of the *useful* streams leaving the process divided by the enthalpy of *all* streams entering the process. In short, it counts the Btu's recovered in any form as a fraction of the Btu's put in.

The second law of thermodynamics makes two statements:

(1)   The quality, or the inherent capacity to cause change, of the input streams is important.

(2)   This quality can be degraded or destroyed by the steps in practical processes.

In various ways thermodynamicists have struggled to quantify the inherent potential to cause change relative to the environment of any material at a given set of conditions. Various definitions of these terms will be discussed later. For the moment the term "available energy" will be used to describe this characteristic. The numbers in the "second law" column of Table 2-1 represent the ratio of the available energy contained in the products of a

Table 2-1

Comparison of First and Second Law
Process Efficiencies (%)

Unit operation (or process)	First law	Second law
Residential heater (fuel)	60	9
Domestic water heater (fuel)	40	2–3
High-pressure steam boiler	90	50
Tobacco dryer (fuel)	40	4
Coal gasification, high Btu	55	46
Petroleum refining	~90	10
Steam-heated reboiler	~100	40
Blast furnace	76	46

process to the available energy in all input streams of that process. Thus, it counts not only the Btu's lost in waste streams from the process but also the degradation of the quality of the Btu's in the product streams which has occurred as a result of the process.

In general the second law efficiencies are much lower than those calculated by a simple enthalpy balance. In the terms of thermodynamics, the irreversibilities of the process have destroyed some of the available energy of the input materials, so that the products have a lower capability to cause change than the feed streams.

Let us consider some of the examples in the table. The residential heater and domestic water heater have relatively low first law efficiencies. These efficiencies represent the fraction of the primary energy source that is captured in the water or comfort-heating medium. The second law efficiency for these processes is even lower because the Btus are captured at a very low temperature (somewhere between 150 and 220°F). Thus, the Btu's have been degraded from the flame temperature (or the pure-work capability of electricity) to a temperature level that can do little more for us than alleviate cold and wash clothes. The same is true of the tobacco dryer, in which heat is used to evaporate water at a very low temperature.

If we consider the high-pressure steam boiler and the coal gasification plant, a somewhat different story appears. The second law efficiencies are much higher because the product obtained from these operations still has a great deal of potential to do work. The high-pressure steam can be put through a turbogenerator to produce electricity and the exhaust steam used to provide low-level process heat or building heat. In the case of coal gasification, the product is another useful fuel that is capable of doing all the things that coal could do in a more easily distributed, less polluting way. The second law efficiency is not 100% in this case because the amount of gas recovered contains only about half of the total Btu's contained in the coal fed to the plant.

Two observations can be drawn from this analysis. The first is that, for a number of processes or operations which consume a large amount of energy, even a very small improvement in the second law efficiency of the process will have a large impact on the amount of fuel consumed.

The second observation is that the systems used to provide our energy needs are inappropriate in some cases. To correct some of these, efforts are being made to develop a heat pump for domestic heating and hot water uses. These machines could cut available energy use by a factor of 10. On the other hand, there are also developments aimed at capturing "free" solar energy to improve the economics of our present inefficient systems. There is a price in capital costs to utilize the "free" solar energy with conventional, low-efficiency technology.

I. Definitions

A. Available Energy, Availability, and Exergy

This book is not a text on thermodynamics. At the same time, it must provide a basic understanding of the application of thermodynamic principles to industrial processes. The objectives of this section of the book will be to do this and also to sort through the variety of terms used to describe approximately the same principles, so that the reader will be equipped to understand the literature and to study further.

The term "capacity to cause change" has already been introduced. Before advancing some quantitative definitions of the appropriate thermodynamic terms, one qualitative example may help to clarify the varying quality of energy sources. If the process objective is to obtain a cup of tea and the designer is equipped with a teabag, a cup, a kettle of water, and two hot-air streams, one at 250°F and one at 160°F, perhaps the differences in "capacity to cause change" can be demonstrated. Assuming the only energy source available to boil the water for the tea is one of the hot-air streams, the choice is simple. Any engineer knows that he is not going to boil water with 160°F air no matter what its price or how many Btu's are available. On the other hand, given enough insulation and patience he can get the water to boil with 250°F air. In thermodynamic terms this difference in capability (quality) is defined as the difference in the available energy of the two streams. Please note that the 160°F air stream does have the capability to heat a house in winter and keep the occupants comfortable. Obviously, the 250°F air stream also has this capability, but it would be wasteful and possibly dangerous to use it for this purpose. The point is that if the available energy of utility streams is matched to the needs of the tasks through the most efficient process we can afford, the utilization of fuel in the system will be optimized.

An interesting example was given by Berg.[2] In an industrial plant application, heat was used to cure resin coatings at the rate of 12 MBtu/h. Research showed that ultraviolet radiation, because of its shorter wavelength, was better suited to curing the resin than the broad-spectrum energy supplied by the heat. Changing to ultraviolet light cut direct energy needs to 0.15 MBtu/h of electricity. Correcting for typical fuel–power efficiencies, the fuel requirement of the light was still only 0.5 MBtu/h. This remarkable change came from simply matching the correct form of energy to the job requirement: the first principle of efficient energy use.

A favorite term of thermodynamicists is the "reversible process." The Carnots (Lazare and his son Sadi) first introduced the concept by observing that the maximum work could be obtained from a process only when the actuating force differences were infinitesimally small.[3] Since the driving forces under such circumstances can be easily reversed with essentially no impact on the environment, the name "reversible" stuck. The Carnots considered mechanical- and temperature-driven systems, and J. Willard Gibbs extended the idea to chemical concentration differences. In short, Gibbs pointed out that, theoretically, work could be obtained by the diffusion of chemical components from one stream to another across infinitesimal concentration differences as well as from interposing heat or mechanical engines between temperature or force differences.

We can conclude, then, that any material whose temperature, pressure, potential energy, velocity, or composition differs from the general environment has the capacity to cause change if we can devise a process or machine to utilize it. This is the available energy of the material. The maximum amount of the desired change will be obtained if a "reversible" process is devised to do the job.

Available energy (denoted by the symbol A) is known by many other names in the literature, some of which have subtly different meanings. Keenan's "availability" and terms involving "work" (available work, ideal work, etc.) originated primarily from mechanical engineering and thermomechanical analysis. These are often perceived as being limited to processes involving mechanical work only, which causes some confusion. However, if one defines the "dead state" (the environment) in terms of composition as well as temperature and pressure, these terms can serve essentially the same purpose as "available energy." The same is true of "exergy," primarily a European term. Evans's "essergy"[4] specifically includes the chemical terms

Table 2-2

Connections among Essergy, Availability, Exergy, and Free Energy[a]

Name	Function	Comments
Essergy	$E + P_0V - T_0S - \sum_i \mu_{i0} N_i$	This function was formulated for the special case of an existing medium in 1878 (by Gibbs) and in general in 1962. Its name was changed from "available energy" to "exergy" in 1963, and from "exergy" to "essergy" (i.e., "*ess*ence of en*ergy*") in 1968 by Evans.
Availability	$E + P_0V - T_0S$ $- (E_0 + P_0V_0 - T_0S_0)$	Formulated by Keenan in 1941, this function is a special case of the essergy function.
Exergy	$E + PV - T_0S$ $- (E_0 + P_0V_0 - T_0S_0)$	Introduced by Darrieus (1930) and Keenan (1932), this function (which Keenan has called the "availability in steady flow") was given the name "exergy" by Rant in 1956. This function is a special case of essergy.
Free energy	Helmholtz: $E - TS$ Gibbs: $E + PV - TS$	The functions $E - TS$ and $E + PV - TS$ were introduced by von Helmholtz and Gibbs (1873). These two functions are Legendre transforms of energy that were shown by Gibbs to yield useful alternate criteria of equilibrium. As measures of the potential work of systems, these two functions represent special cases of the essergy function.

[a] The terminology used is defined on p. 23.

in its definition and therefore most precisely represents the full potential of a material not in equilibrium with its environment. The term is not widely used in the literature or in textbooks to date. In Table 2-2 a more detailed comparison of definitions is given.

Consequently, in reading the literature one has to take the time to ascertain the particular author's meaning for the terminology he uses. Many times it will boil down to a common practical interpretation, but not always. One example occurs when the improvement of heating, ventilation, and air conditioning (HVAC) systems is being considered. Here, the condensation and evaporation of water in the air introduces composition changes that must be accounted for.

Quantitative definitions of A must begin with the definition of that dreaded term "entropy" (S). As derived in a number of texts on thermodynamics,[5] entropy is defined as

$$dS = dQ/T, \tag{2-1}$$

$$\Delta S = S_2 - S_1 = \int_1^2 dQ/T, \tag{2-2}$$

and, for a reversible process,

$$\Delta S = 0 = \int_1^2 dQ/T. \tag{2-3}$$

In the reversible case the heat Q must be transferred through infinitesimal temperature differences to the surroundings and might be termed dQ_{rev}, the "reversible" heat exchanged.

In the real world no processes are reversible. Not only are infinitesimal driving forces uneconomic (generally, although some cryogenic processes come close), but they are also very hard to control. The ideal of a reversible process is worth considering, however, because it represents the best that can be done by *any* process in accomplishing a given change. Another way to define the second law efficiencies given in Table 2-1 would be the ratio of the reversible ΔA to the actual ΔA for any process, i.e., the best possible amount of work divided by the actual work in a process that requires work. Since efficiency terms per se do not help in improving processes, the debate over which efficiency definition to use will be left to others.

Inherent in the preceding discussion is that the value of S for a material, like the values of H, E, and other extensive properties of the material, depends only on the initial and final states of the material. These properties are independent of the process path. A reversible and irreversible process between the same initial and final states for a material will end up with the same *values* for H, S, A, etc., but the work requirement (or production) will be different for each process path, so the fuel requirements for the process will also differ. Simple evidence of this can be obtained from the steam tables. The property values for 600-psig, 750°F steam are constant whether it is produced in a boiler from fuel or as turbine exhaust from higher-pressure steam, but less fuel needs to be burned in the latter case after the fuel attributed to the power generated is subtracted.

Many investigators (see Table 2-2) have contributed to a definition for available energy which applies in the general case in which work, heat, and mass (composition) come to equilibrium with an environment and cause change in doing so. For a closed system, A_{cs} for a state higher than a "dead state" defined by temperature T_0, pressure P_0, and composition μ_{i0} is

$$A_{cs} = E - E_0 + P_0(V - V_0) - T_0(S - S_0) + \sum_i N_i(\mu_i - \mu_{i0}), \tag{2-4}$$

where E, V, S, and μ_i represent the internal energy, volume, and entropy of the system and the chemical potential of each component, respectively, E_0, V_0, S_0, and μ_{i0} represent these quantities for the "dead" (or equilibrium) state, and N_i is the number of moles of species i.

In general the term E should include all elements of the energy system:

gravitational (potential), kinetic, magnetic, and electric energy, etc. In industrial processes these are often neglected and only the thermal–mechanical elements are considered.

The chemical available energy (or diffusion) term is significant only for chemically reactive processes (such as combustion). For other processes semipermeable membranes would be required to recover work by diffusion. At this writing such devices do not exist in industry. As a result the chemical term is often dropped in practical analyses.

For a flow system the flow work $(P - P_0)V$ must be added to Eq. (2-4). This results in

$$A = E - E_0 + PV - P_0V_0 - T_0(S - S_0) + \sum_i N_i(\mu_i - \mu_{i0}). \quad (2\text{-}5)$$

Noting that $H = E + PV$, Eq. (2-5) becomes

$$A = H - H_0 - T_0(S - S_0) + \sum_i N_i(\mu_i - \mu_{i0}). \quad (2\text{-}6)$$

Eliminating chemical terms, where appropriate, leaves

$$A = (H - H_0) - T_0(S - S_0), \quad (2\text{-}7)$$

which is useful in many practical problems.

It should be noted that if only changes in A for two states of the same composition are sought, the dead-state terms H_0 and S_0 drop out, and

$$\Delta A = \Delta H - T_0 \, \Delta S, \quad (2\text{-}8)$$

where

$$\Delta H = H_2 - H_1 \quad \text{and} \quad \Delta S = S_2 - S_1. \quad (2\text{-}9)$$

This has significance in simplifying many calculations, e.g., energy flow in steam systems.

To avoid confusion later note that A, the available energy of a material, is not necessarily the same as the Gibbs free energy (generally denoted by G) of the material. By definition,[6]

$$\Delta G = \Delta H - \Delta(TS). \quad (2\text{-}10)$$

For an isothermal (constant-temperature) process at the temperature T_0 of the environment, ΔG reduces to

$$\Delta G_{T_0} = \Delta H - T_0 \, \Delta S, \quad (2\text{-}11)$$

which is identical with ΔA. For the special case in which a process occurs at constant temperature, ΔG is a measure of the work that can be extracted from an *isothermal change of state,* but ΔA is completely general — it measures the work extractable (or required) from *any change of state.*

B. Looking at a Process Backward

We have talked about recovering work or other "capacity to cause change" in a process. Since processes do not always occur spontaneously, available energy must be put in at some point in the process to drive it. From the definition of a reversible process, it should be obvious that ΔA_{rev} measures the *minimum input* to a process that requires work as well as the *maximum output* of a process that delivers work.

In real processes the ΔA required to make the process occur, of course, is larger than ΔA_{min}. Available energy is lost or degraded because of process irreversibilities. Here is where the differences between the first and second laws become apparent; all of the Btu's entering the process can be accounted for, but their "capacity to cause change" is reduced. Consequently, available energy is *consumed* in the process. This consumption is what we pay for in fuel costs and what we must strive to minimize. In the literature the consumption of available energy is sometimes termed "lost work." If "work" is interpreted in the general sense discussed earlier, this terminology is appropriate.

C. Available Energy Balances

Steady-state balances for available energy can be constructed in the same way as the steady-state enthalpy (energy) balance is, with one exception. Whereas the enthalpy is conserved, available energy can be destroyed or degraded, so its balance equation must contain destruction (A) and production (A_{prod}) terms. Therefore, the total balance can be written

$$A_{in} = A_{out} + (A_{prod} - A_{dest}).\qquad(2\text{-}12)$$

Since the available energy is a state function, as are enthalpy and entropy, in steady-state cases Eq. (2-7) can be used to calculate the available energy of any component if H and S are known and an appropriate ambient temperature is established. For example, in a steam turbine the available energy of the inlet steam and the outlet steam can be calculated from the steam tables. At any given turbine efficiency, the work produced by the turbine can be calculated from the Molliere chart, and the available energy destroyed by the inefficiency of the turbine can also be calculated. The sum of these terms must balance. The change in the available energy of the steam in passing through the turbine is then equal to

$$\Delta A = W + A_{dest},\qquad(2\text{-}13)$$

where W is the actual work produced by the turbine and A_{dest} the available energy destroyed.

A key point in the manipulation of these properties is that the base point of the data for all materials must be consistent if we are to have consistent *absolute values* for H, A, S, etc. On the other hand, if we are dealing only with *changes* in properties, the base conditions cancel out. This is particularly convenient when dealing with steam, air, and other common items.

D. Energy Change In Terms of Work

Available energy changes are often described in terms of work. With the sign conventions to be established in Fig. 7-2, where work done *to* a process is negative, the ideal work has the opposite sign from the ideal available energy, ΔA_i. These relationships can be simply derived.

The first law energy balance can be written

$$\Delta H + KE + PE = \sum Q - W_{act}, \qquad (2\text{-}14)$$

where KE and PE are the kinetic and potential energy terms, which will be dropped forthwith to simplify the example. Heat must be exchanged with the surroundings, resulting in an entropy change of

$$\Delta S_{sur} = \sum Q_{sur}/T_0. \qquad (2\text{-}15)$$

In a *reversible* process the total entropy change must be zero, i.e.,

$$\Delta S_{tot} = 0 = \Delta S_{sur} + \Delta S_{proc}, \qquad (2\text{-}16a)$$

or

$$-\Delta S_{sur} = \Delta S_{proc}. \qquad (2\text{-}16b)$$

Substituting these relationships into Eq. (2-14), ΔH becomes

$$\Delta H = -T_0 \, \Delta S_{sur} - W_i \qquad (2\text{-}17)$$

and then

$$\Delta H = T_0 \, \Delta S_{proc} - W_i, \qquad (2\text{-}18)$$

or

$$W_i = T_0 \, \Delta S_{proc} - \Delta H. \qquad (2\text{-}19)$$

Referring back to the derivation of ΔA and neglecting the chemical and corresponding energy terms [see Eq. (2-8)], the expressions translate to

$$-W_i = \Delta A_i.$$

Again, parallel to the development of available energy balances,

$$W_{lost} = W_i - W_{act}. \qquad (2\text{-}20)$$

In a work-producing process, this is apparent without concern about the sign of any term. However, in a process that requires work, W_{act} is numerically greater (more negative) than W_i, so W_{lost} is still a positive quantity. It may help to think of W_{lost} as work dissipated *on* the surroundings and, hence, a positive number at all times.

Substituting from Eqs. (2-14) and (2-19), the lost work becomes

$$W_{lost} = T_0\,\Delta S_{proc} - \Delta H - \left(\sum Q - \Delta H\right), \qquad (2\text{-}21)$$

or

$$W_{lost} = T_0\,\Delta S_{proc} - \sum Q. \qquad (2\text{-}22)$$

Note that

$$Q_{sur} = -\sum Q = T_0\,\Delta S_{sur}; \qquad (2\text{-}23)$$

therefore,

$$W_{lost} = T_0\,\Delta S_{proc} + T_0\,\Delta S_{sur} = T_0\,\Delta S_{tot}. \qquad (2\text{-}24)$$

The value of these expressions is that they provide ways to calculate W_{lost} directly for a process segment if the system properties and flow rates are known, where W_{lost} corresponds to the net $A_{prod} - A_{dest}$ term in the available energy balance [Eq. (2-12)].

II. Available Energy and Fuel

It has been stressed that fuel is the usual source of available energy in the industrial environment. One way or another, this is what we pay for (with associated charges for the capital invested in the combustion–energy transport system). However, the combustion process itself is highly irreversible and therefore inefficient.

Gyftopoulos[7] demonstrated the inefficiency of combustion in terms we can now understand. Consider a hypothetical fuel CH_2. This fuel has a lower heating value of 280,000 Btu/lb · mole (the higher heating value is 298,000 Btu/lb · mole) calculated from the enthalpy change of the following stoichiometric reaction:

$$CH_2 + \tfrac{3}{2}O_2 + 5.65N_2 \rightarrow CO_2 + H_2O + 5.65N_2. \qquad (2\text{-}25)$$

Let us accept Gyftopoulos's calculation that the available energy change for the reaction at $T_0 = 55°F$ would be about 290,000 Btu/lb · mole (point a in Fig. 2-1; some 97.6% of the higher heating value) and concentrate on

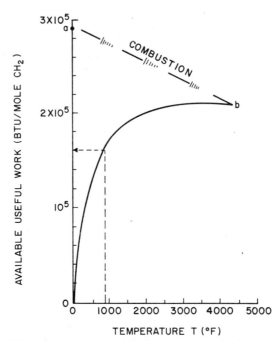

Fig. 2-1   Available useful work from the hydrocarbon oxidation process: ——, from products of combustion and material heated by them at temperature T; ---, from 1500-psia superheated steam. From Reference 7.

finding a reversible process by which this energy can be recovered. One such process would have four steps.

(1)   The oxidation is carried out in a reversible fuel cell, at some temperature T_a, which delivers electrical work to the surroundings.

(2)   The products of the fuel-cell process are cooled from T_a to the temperature of the environment T_0 as they provide heat to Carnot engines which produce further work.

(3)   Each of the products CO_2, H_2O, and N_2 is separated from the mixture reversibly by means of a semipermeable membrane and expanded reversibly and isothermally at T_0 in an engine until it attains a pressure equal to the partial pressure of that constituent in the atmosphere.

(4)   Each molecular species is introduced reversibly into the atmosphere through a semipermeable membrane. The term "semipermeable membrane" refers to a device which is impermeable to all molecular species except one.

The final state after Step (4) corresponds to zero available useful work at atmospheric temperature (55°F).

Practical devices to accomplish these steps do not yet exist. In the real world the fuel is fired in a furnace or boiler and then the heat in the flue gas (products of combustion) is recovered by heat transfer. Figure 2-1 shows the impact of eliminating the membranes, Carnot engines, and fuel cells from the process. Assuming the stoichiometric ratio of air to fuel, the flame temperature should be 4300°F (point b in Fig. 2-1). Because there is no fuel cell delivering work directly from the combustion reaction, about 88,000 Btu, or 27% of the available energy of the fuel, is immediately lost. The maximum available energy that can be recovered from this amount of 4300°F flue gas, *even reversibly,* is only about 200,000 Btu.

About the best practical process used in recovering this available energy is the production of high-pressure, superheated steam. The dashed line in Fig. 2-1 shows the amount of useful energy recovered if *all* of the flue gas is used to produce 950°F stream, e.g., superheated 1500-psig steam. The available energy captured in the steam would be about 160,000 Btu, or 55% of that in the fuel. Please note that more irreversibilities will be encountered as the steam is used to cause change because of machinery friction, temperature driving forces, etc.

To summarize, even if we could recover all of the available energy at 950°F, we would have only about half of the available energy of the fuel (but almost all of the Btu's in the fuel). The remainder is lost because there is no process to deal with the combustion reaction reversibly, and because there is no practical way to utilize the heat at temperatures higher than 950°F. The real situation in a 1500-psig boiler is even worse in that a large share of heat is transferred at 625°F (the saturation temperature of 1500-psig steam). At this temperature only 130,000 Btu would be available and the recovery would be reduced to 45%.

In chemically reacting systems such as combustion, ΔA for a reaction can be estimated from the differences in the Gibbs free energy of formation between reactants and products. As we shall discuss in later chapters, these values are available for many materials. Thus, the available energy change for a reaction (the reaction work, as it were) can be calculated from

$$\Delta A_R = \sum_i G_f^0 \text{ (reactants)} - \sum_i G_f^0 \text{ (products)}. \qquad (2\text{-}26)$$

If ΔA_R is positive, energy is released in the reaction (the reaction is exothermic), such as occurs during combustion. If ΔA_R is negative, work input is required to drive the reaction (it is endothermic), such as occurs during steam cracking. The methods of handling these different chemical energy effects are the same.

As a further example of how this approach can be applied, consider the data in Table 2-3. Reistad[8] did a simple comparison of where energy (Btu's) and availble energy is lost in a coal-fired, steam-electric generating plant.

Table 2-3

Energy and Available Energy Losses in a Coal-Fired
Steam Electric Generating Plant

Component	Energy loss (% plant input)	Available energy loss (% plant input)
Steam generator	9	49
Turbines	0	4
Condenser	47	1.5
Heaters	0	1.0
Miscellaneous	3	5.5
Total	59	61

The total percentage losses were not much different, but the distribution was markedly so. Thus, the direction given the engineer charged with improving the process is also markedly different. This difference in viewpoint is the key element in obtaining profitable energy saving ideas. Engineers must concentrate on recovering energy that is useful to their industrial processes. In the end, recovered energy must save fuel or its equivalent to be valuable. Much more will be said about this in Chapter 9.

Summary

In this book the available energy A and ΔA in one form or another will be the primary variable rather than enthalpy or "heat." In a thermomechanical process, ΔA and various "work" terms will be interchangeable. These terms will be used to focus on the quality aspects of energy and to identify *losses* in processing systems which offer potential for reduction. This new viewpoint will be trained to focus on practical profit (cost reduction) opportunities through a variety of approaches.

Notes to Chapter 2

1. R. Gaggioli, Second Law Analysis for Process and Energy Engineering. Paper presented at the AIChE Meeting, November, 1980.

2. C. A. Berg, A technical basis for energy conservation. *Mechanical Engineering* **96**, 30 (1974).

3. M. Sussman, "Availability (Exergy) Analysis," Milliken, Boston, Massachusetts, 1980.

4. R. B. Evans, "A Proof that Essergy is the Only Consistent Measure of Potential Work in

Chemical Systems." Ph.D. Dissertation, Department of Engineering, New Hampshire, Dartmouth College, 1969.

5. B. Dodge, "Chemical Engineering Thermodynamics," p. 65. McGraw-Hill, New York, 1944.

6. B. Dodge, "Chemical Engineering Thermodynamics," p. 86. McGraw-Hill, New York, 1944.

7. E. P. Gyftopolous, L. J. Lazaridis, and T. F. Widmer, "Potential Fuel Effectiveness in Industry," p. 11. Ballinger, Cambridge, Massachusetts, 1974.

8. G. M. Reistad, Available Energy Conversion and Utilization in the U.S., *Journal of Engineering for Power* **97,** 429–434 (1975).

3

Thermodynamics and Economics
Part I

Introduction

In this chapter the fundamentals of pricing various utility streams will be discussed. The purpose will be to give the reader an understanding of the value of various utility streams in a process plant. The principles by which plant utility streams can be priced correctly will be described, the objective being to foster the energy efficiency of the site. The term "thermoeconomics" will be avoided in these discussions because it is reserved for a more specific definition. In Chapter 11 the comprehensive "thermoeconomics" systems proposed by investigators such as Gaggioli, Tribus, and Evans will be discussed briefly.

I. General Considerations

In Chapter 2 some sources of energy were described as more useful and therefore worth more to a plant operator than others. In our terminology, they contained more "available energy" (exergy, essergy, availability) than other sources of energy. The goal of the energy conservation engineer is to use as much as possible of the contained available energy in any utility stream to minimize the energy cost of the process. In addition, the engineer wants to price any surplus energy export streams correctly to ensure a market and maximize the profitability of the operation.

A. The Qualitative Picture

From the principles described in Chapter 2, we recognize the qualitative differences between energy streams based on what each stream can accomplish for us. Figure 3-1 presents a qualitative picture, or ladder, of this concept. A building can be heated by any of the utility streams listed. Water

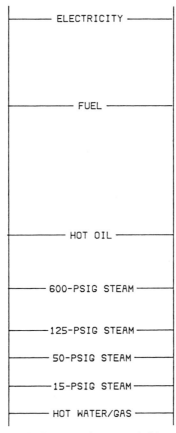

Fig. 3-1   Relative energy ladder.

can be boiled using any of the utility streams except for the lowest ranked. However, only fuel or electricity can supply energy at a high enough level to heat the reboiler on a crude oil pipestill to 800°F. In addition, considerably more horsepower can be derived from electricity than from the same number of Btu's of fuel. Based on the specific properties and conditions for each utility stream, a quantitative value for the available energy above a fixed dead state (environment) can be assigned to each of the utilities pictured.

B. Relation of Available Energy Content to Price

How do we relate this inherent characteristic of each utility stream to an economic price? Fundamentally, the prices must be related to the real prices paid for the primary sources of available energy. These usually consist of fuel and electricity, but circumstances are different in each plant.

Generally three combinations of purchased or self-generated utilities exist in the process industry. The first is the totally integrated energy system, which generates both steam and power (as economic) from self-generated and/or purchased fuel. Plants using this system may also import or export power, steam, or fuel, depending on the balances of the moment. Integrated steam power systems may not be complete at each site, yet this remains the objective, because a balanced system provides the best utilization of available energy. The second type of plant burns fuel on site but imports all power from a utility source. Under these circumstances, available energy can be tied directly to the price of purchased power as well as fuel. Difficulties sometimes arise over increments of demand charge and between average and incremental costs. The third type of plant is one that buys all of its utilities from an outside company. Here, ensuring that these utilities are priced properly becomes the primary concern.

C. Steam Pricing

Wherever possible, utility prices should be set to foster energy efficiency. This means each plant should strive to reduce the net cost to the operation for its required energy input. If utilities are priced correctly, the lowest-value utility will be used for each process job, and most investments that move the plant closer to its ultimate energy efficiency will be justified in times of normal capital availability. Specifically, steam should be priced to

(1) promote steam conservation by process optimization and process control,

(2) encourage maintenance to reduce losses and waste,

(3) encourage production at the highest possible pressure and consumption at the lowest possible pressure,

(4) discourage the use of condensing turbines or inefficient back-pressure turbines,

(5) encourage interunit heat integration, and

(6) encourage any other plant modernization projects that increase the second law efficiency.

D. Integrated Plant Energy Systems

Although this system is the most complex, the principles of pricing steam utilities can be most readily understood in this context. Consider the simple

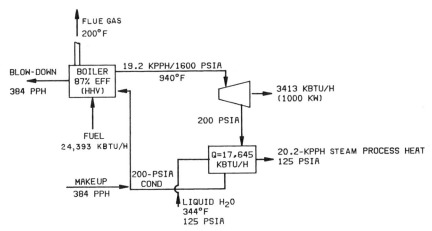

Fig. 3-2   A simple steam power cycle with a balanced cogeneration system.

steam power cycle shown in Fig. 3-2. The balanced cogeneration system shown produces about 1 MW of electricity and something over 17 MBtu/h of process heat in the form of 125-psia steam from 1600-psia, 940°F steam raised in a very efficient boiler system. The 87% overall boiler efficiency includes the steam consumption of the boiler itself and really represents a 92% first law boiler efficiency and approximately 0.06 lb of steam consumed in boiler auxiliaries for each pound of steam exported from the boiler complex to the rest of the plant. In addition, a relatively low blow-down rate of 2% is included.

From this well-balanced system, let us consider the case in which the process heat demand is reduced by 50% owing to the shutdown of a consuming unit, but power demand is constant. This example will help us to evaluate a fair cost for the exhaust steam produced from the turbine. The rebalanced case is shown in Fig. 3-3, and it requires the use of a condensing steam turbine that exhausts at an absolute pressure of 1 psia so that power output can be maintained in the face of reduced process heating demand. Comparing the two diagrams, we see that high-pressure steam production is reduced from 19,200 to 13,550 lb/h, of which 3950 lb/h go through the condensing portion of the turbine. Fuel consumption is reduced from 24.4 to 18.5 MBtu/h. This represents a saving of approximately 5.9 MBtu/h compared with a reduction in process heating demand of about 8.8 MBtu/h. The available energy savings are less than the enthalpy demand reduction because the overall efficiency of the system changes. If the fuel is assumed to cost $5/MBtu, savings in fuel cost will be $3.34/MBtu of reduced process heating demand. Conversely, the cost of supplying an additional process

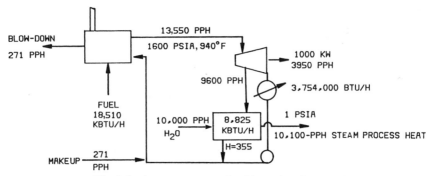

Fig. 3-3 A steam power cycle with a reduced process load.

heating demand for the case pictured in Fig. 3-3 is also $3.34/MBtu, assuming that the investment to provide all streams is already in place. The increase in system efficiency resulting from full utilization of the cogeneration potential of the plant makes this increment very attractive.

Note that this system provides the first evidence of interaction among various utility streams. Because the plant is forced away from its optimum configuration by changes in loads, it cannot recover the full fuel price for those heat increments no longer required. This concept of interaction will be discussed further in Chapter 6.

E. The Impact of Turbine Exhaust Pressure

These calculations were repeated for various turbine exhaust pressures in the range from 140 to 600 psia, *still assuming that the process required only 125-psig steam to meet its needs.* The results are shown in Table 3-1. From a thermodynamic viewpoint, the use of high-pressure exhaust steam to supply the necessary process steam at 125 psia represents the introduction of an additional inefficiency compared with the base case. Larger temperature differences in the 125-psia steam generator reduce the surface area required but increase the thermodynamic inefficiency of the heat exchange. This change in the thermodynamic inefficiency of the system is reflected in boiler-fuel savings for the same absolute change in process heating duties. The cost of producing the extra process heating load is lowest for the most efficient system, i.e., the one with the smallest temperature driving force in the steam generator. It is almost 50% higher for the least efficient system (highest ΔT) considered.

Clearly, calculating back to the fuel impact of any change in system energy demand represents an accurate method of evaluating the cost and/or saving

Table 3-1

The Impact of Change in Process Heat Load on the Amount of Fuel Fired at Constant Power Generation

Exhaust pressure (psia)	Heat load (MBtu/h)	Total high-pressure steam (pph)[a]	Steam to condensing (pph)	Fuel fired (MBtu/h)	ΔFuel (MBtu/h)	ΔFuel/ΔQ[b] ($/MBtu)
140	8.406	12,978	4407	18.124	—	—
	17.226	18,310	0	23.903	5.779	3.28
200	8.825	13,550	3950	18.510	—	—
	17.645	19,200	0	24.393	5.883	3.34
400	19.956	24,132	2417	29.730	—	—
	28.776	31,312	0	37.298	7.568	4.29
600	31.777	36,887	1714	42.695	—	—
	40.597	44,908	0	51.016	8.321	4.72

[a] Pounds per hour.
[b] The assumed fuel cost is $5/MBtu.

associated with adding or subtracting an energy demand from the system. This ties the cost of any utility directly to the real cost of fuel at the site in question. Since fuel is a primary source of available energy, the cost of available energy in any stream can be calculated directly.

This approach to pricing intermediate utility streams has two drawbacks. The first is that prices fluctuate with the efficiency designed into the increment of utility added or subtracted. In the example discussed, the price of 125-psia steam would depend on its source and could not be generalized. The second problem is far greater. In practical plant systems, the difficulty of making the calculation becomes great because of the large potential for interaction among the many system components. In a large plant there would be many opportunities to reoptimize the steam power system in the face of the demand change considered in this example. The interactions of all these opportunities would have to be known before the best configuration could be selected and the lowest cost calculated for the demand change. In addition, the optimum configuration might change seasonally or diurnally or be a function of plant throughput, feed slate, or external considerations. It is possible to analyze one's steam power system and generate the necessary computer programs to do this; however, it is far from the norm in operating facilities. In addition, the operation becomes so complex that people interested in developing projects to improve energy efficiency become confused and lose momentum.

The following conclusions can be drawn from the exercise so far.

(1)   The cost of changes in available energy demand can be related directly to fuel in an integrated steam power system.

(2)   These costs are a function of the efficiency of the system being considered and not a constant times the price of fuel.

(3)   Except for very simple systems, the complexity of this calculation imposes large difficulties in making all but the most major decisions.

As a result, engineers involved in energy conservation, process designers, plant managers, and even accountants look for simplified ways to aproximate the costs of all utilities in a specific site situation. The objectives of the pricing system do not change. Whatever simplified system is adopted must be tested to be sure that the resulting prices work to enhance the energy efficiency rather than to confuse. With an effective system, the complexities of multiple parallel decisions about energy utilization and project development can be simplified. However, it must be recognized that any system of pricing based on the properties of steam in a header will not represent the specifics of each turbine, but rather the average of all sources of steam in the particular header. Variations in machine efficiencies, letdown, station performance, etc., could mask the right answer for any proposed facilities change. It is recommended that for major investment and pricing decisions the specific impact on fuel be calculated whenever possible as a check on the accuracy of the approximate methods.

II. A Systematic Approach to Steam Pricing

A. Enthalpy-Based Prices Are Incorrect

A common approach to pricing steam at intermediate pressure levels is to consider the total enthalpy of the stream. The total cost of steam at the generation pressure is generally well known. An example of how these costs are calculated will be shown later in this chapter. In the enthalpy pricing approach, the ratio of the enthalpy of the steam at lower pressures to the enthalpy of the steam at generation pressure is used to reduce the cost of steam at generation pressure to the price for the lower-pressure steam. For the pressure levels discussed in the previous example, the enthalpy ratios are shown in Table 3-2. By inspection, the enthalpy ratios in this table show very small changes between the pressure levels considered if steam prices are determined by enthalpy ratios. There will be high driving forces for the installation of back-pressure turbines, because the exhaust steam has nearly the same value as the inlet steam, resulting in very cheap prices for the power generated. Also, the recovery of waste heat to generate medium-pressure steam will be encouraged because of the high price of medium-pressure

Table 3-2

Table of Property Values for Steam at Various Pressures[a]

Pressure (psia)	Temperature[b] (°F)	Enthalpy (Btu/lb)	Enthalpy ratio	Entropy (Btu/lb °R)	Availability (Btu/lb)	Availability ratio
1515	950	1459.2	1.0	1.5790	608.9	1.00
615	740	1373.2	0.941	1.6040	513.7	0.843
315	550	1284.7	0.880	1.5920	427.4	0.702
190	450	1240.0	0.850	1.5988	379.0	0.622
140	400	1221.1	0.834	1.6087	351.3	0.577
125	Sat	1191.0	0.816	1.5845	337.9	0.555
65	Sat	1179.1	0.808	1.6374	297.3	0.488
35	Sat	1167.1	0.800	1.6870	258.5	0.425
15	Sat	1150.4	0.788	1.7566	204.2	0.335
10	Sat	1143.3	0.784	1.7876	180.4	0.296
5	Sat	1131.1	0.775	1.8441	137.7	0.226
2	Sat	1116.2	0.765	1.9200	81.8	0.134
0.698	Sat	1100.9	0.754	2.0087	18.6	0.030
0.507	Sat	1096.6	0.752	2.0360	−0.4	0

[a] Based on 80°F.
[b] Sat stands for saturated.

steam. However, there will be little chance to justify the extra surface area required in applications in which lower-pressure steam is thermodynamically possible.

The theoretical inappropriateness of using the enthalpy ratios to set steam prices is shown by the curve marked "enthalpy" in Fig. 3-4. The zero pressure intercept for the enthalpy ratio is about 0.75. If high-pressure superheated steam costs $8/1000 lb, it would be absurd to assign a value of $6/1000 lb to steam at zero pressure and 80°F. Such a utility would be useful for heating greenhouses, but little else. No one would pay $6/1000 lb for it because cheaper, more generally useful alternatives would be available. Thus, the enthalpy ratio is not an appropriate basis for pricing steam at pressures lower than that in the boiler.

B. Available Energy Prices Are More Correct

A more correct thermodynamic approach would be to price the steam according to the ratio of contained available energy to available energy in the steam at generation pressure. This approach places an equal value on the available energy content of the steam at all pressure levels. The same method of calculation is used, namely, the price of steam at lower pressures and

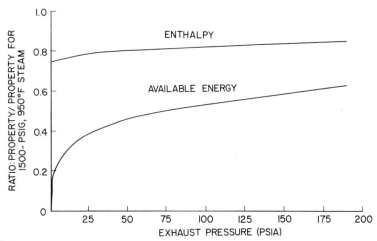

Fig. 3-4   Relative enthalpy and available energy values of exhaust steam versus power steam.

temperatures is equal to the available energy content at the lower level divided by the available energy content at the generation pressure and temperature multiplied by the cost at generation pressure. These results are plotted as the available energy curve in Fig. 3-4, and several values are listed in Table 3-2. The data satisfy the theoretical intent that steam has zero value as it reaches ambient conditions and show a much larger difference in prices between pressure levels. From a practical standpoint, the result of the available energy pricing system is to make critical the justification for efficient back-pressure steam turbines. In addition, the value of steam recovered in a waste heat boiler is lower, making their use more difficult to justify. In return there is a much greater incentive for using the correct pressure level of steam to match the process requirements and greater incentives to improve overall system efficiency.

Gaggioli and Wepfer expanded on the utilization of an available-energy-based pricing system for utilities in some detail.[1] Their observations are pertinent only to the specific situations in the paper plant they studied, but the analysis of a more complex system is valuable. Another treatment reaching the same conclusion is thus presented: available energy pricing makes sense.

In the general guidance of an energy conservation program, however, the available energy pricing approach is less complex for plants in which the steam power system is in reasonable balance. If special pricing situations are required to force the system toward balance, the available energy pricing formula should be used as a sensitivity test against which projects are evaluated once the long-range goal of a balanced energy system is achieved.

These values will be closer to the correct ones than any artificially contrived system aimed at remedying a steam balance problem. For example, for cases in which a low-pressure steam is being vented, it might be priced at zero value until the system is returned to balance. Under such circumstances incentives for using low-pressure steam would be high and those for recovering waste heat to produce low-pressure steam would be zero. Each project that takes advantage of this temporary high incentive for using low-pressure steam should be checked against the eventual long-term values to be sure that the payout for the required investment has been achieved prior to returning to a normal balance.

Please note that any ratio system for pricing will give different ratios if the conditions of the highest-pressure steam on which it is based are changed. Table 3-2 showed enthalpy and available energy ratios for various steam conditions based on 1515-psia, 950°F steam. These values are plotted in Fig. 3-4. Basing the ratios on 600-psia, 750°F steam would result in markedly different ratios. For example, the ratios for 140-psia, 400°F steam would be 0.883 for enthalpy and 0.686 for available energy on a 600-psia base versus 0.834 and 0.597, respectively, on a 1515-psia, 950°F base.

More sophisticated systems for steam pricing are possible if detailed data on current steam balances are available. A money balance can be created (possibly on a real-time basis) to determine the exact cost of the next pound of steam at given conditions. As discussed on p. 36, this would depend very much on the specific equipment operation and loads of the moment and would require sophisticated computer analysis at all but the simplest plant sites.

Returning to the availability price ratio, adjustments are sometimes required to be sure that back-pressure turbine investments are justifiable. This will be a function of the hurdle rates imposed on energy conservation investments at the plant. The purpose of these adjustments should be aimed at the primary objectives of a sound energy pricing system, namely, to minimize the energy costs at the site.

C. Adjustments to Available Energy Pricing

As shown previously, a large range of values for steam at lower-than-generated pressure can be calculated depending on the concept used for the calculation. The available energy pricing system is the theoretically correct approach. In a practical sense, the energy efficiency engineer wants to reduce fuel costs in his plant to the economic minimum. The plant system of utility values should foster this objective. A three-part test is proposed to check the efficiency of any utility value system.

(1)   Are efficient back-pressure–extraction turbine drivers justified?

(2) Is the extra heat exchange surface required to use lower-pressure steam justified?

(3) Is the recovery of waste heat justified?

As indicated earlier the critical element in using available energy pricing structures is generally the justification of using back-pressure turbines compared with motors. Following Gaggioli and Wepfer,[1] the total cost of steam at the highest (boiler) pressure is calculated by adding the measured variable operating cost to the fixed cost of the boiler house and dividing by the *net production* of steam to consumers. Thus, the average *unit cost* α_{HP} for high-pressure steam in dollars per 1000 pounds is

$$\alpha_{HP} = \frac{\alpha_F F + \alpha_W W + \alpha_C C + R_I K + L}{\text{net steam production}}, \qquad (3\text{-}1)$$

where α_F, α_C, and α_W are the prices of fuel, chemicals, and makeup water, respectively, F, C, and W are the flow rates of these commodities, R_I is the fraction of total capital cost incurred annually for depreciation, maintenance, and return on investment, K is the capital invested in dollars, and L is the cost of labor in dollars.

Boilers generally consume some of the steam they produce in their own fans, pumps, etc., so that the net steam production is lower than the total amount of steam generated. As long as the total flows F, W, and C and the net steam output are known, the calculation of unit steam cost is not complicated by this factor. If all of this data is not available, some correction from gross to net steam production may be necessary. This is generally in the range of 5 to 10%.

As an example, consider a system that produces an average of 600,000 lb/yr of 600-psia, 750°F steam for consumers from three boilers which cost a total of $21 million. The rated capacity of these boilers would be higher than the net production to allow for legal inspections, reserve factors, mechanical outages, and variations in load. If two men per shift ($20,000/man · yr) are required to operate the boilers, maintenance costs are 2% of capital, and return and depreciation amount to 25% of capital, the last two terms (fixed cost) of Eq. (3-1) can be calculated as

$$\text{annual fixed cost} = \frac{(21 \times 10^6)(0.25 + 0.02) + 2(20,000)(4.2)}{600 \times 8760}$$

$$= \$1.11/1000 \text{ lb.}$$

Variable costs are calculated from direct consumptions per unit of production. Assume the following data:

fuel fired per pound of steam (net)	1400 Btu
fuel value	$4.5/MBtu

boiler feed water used per pound of steam (net) 1.1
cost of BF water $0.36/1000 lb
amount of chemicals (net)   0.0001 lb
cost of chemicals $1/lb

$$\text{variable cost} = \left(\frac{1400(4.5)}{10^6} + \frac{1.1(\$0.36)}{10^3} + 0.0001(1) \right) 1000 = \$6.80/1000 \text{ lb}$$

total steam cost = $7.91/1000 lb

The data in Table 3-2 can be used to calculate the appropriate price for this steam when expanded through a turbine to any pressure consistent with the capabilities of the machine. For example, if 140-psia, 400°F steam is chosen, the availability ratios of the 600-psia and 140-psia material are 0.843 and 0.577, respectively. Thus, the value of the 140-psia steam should be

$$\alpha_{140} = (0.577/0.843)\alpha_{600} = 0.684(\$7.91) = \$5.41/1000 \text{ lb.}$$

Whether use of a back-pressure turbine can be justified will depend on its efficiency and the cost of supplying the necessary horsepower by other means. For example, a 500-hp (373-kW) back-pressure turbine would cost about $900,000 versus $360,000 for a motor. To provide a reasonable return (25%) on the incremental investment for the turbine ($540,000), energy costs for the turbine must be about $135,000/yr less than purchased power. This amounts to a difference of about 4¢/kW·h for 8650-h/yr operation. At 100% turbine efficiency, about 22 lb of 600-psia, 740°F steam exhausting at 140 psia would be required to produce 1 kW·h of power. The steam cost of the power would be $0.022($7.91 − $5.41) = $0.055/kW·h. Obviously, puchased power costs must be about 9.5¢/kW·h to justify buying the turbine. At lower turbine efficiencies, the number increases proportionately, e.g., at 70% efficiency the steam cost of power is almost 8¢/kW·h and the required electricity price is 12¢/kW·h.

If steam were priced on an enthalpy basis, the required power costs would be lower. The ratio of steam costs would be 0.886, the steam cost at the outlet $7.01/1000 lb and the power cost 2¢/kW·h. Obviously, justifying the turbine is much easier under this pricing scenario but so is the potential for an erroneous investment decision.

The potential for blending the two approaches to provide a set of steam prices to meet the three objectives with which we started is obvious.

D. Justification of Other Efficiency Steps

Availability values generally provide enough incentive to meet the other two tests outlined previously. For example, if a project is proposed to substitute 35-psia steam for 125-psia steam in a heat exchanger, the cost for

Table 3-3

Effect of Log Mean Temperature Difference (LMTD) on Return for a 5.2-MBtu/h
Steam Reboiler[a]

Process stream temperature (°F)	LMTD for existing 125-psia reboiler (°F)	LMTD for proposed 35-psia reboiler (°F)	Original surface area (ft²)	Incremental surface area (ft²)	Incremental cost (thousands of dollars)	Simple return (%)
210	131	46	580	1070	107	66
220	121	36	628	1483	148.3	47.7
230	111	26	685	2240	224	31.6
240	101	16	753	4000	400	17.7

[a] Based on availability ratio pricing and a cost of $7.91 for 600-psia, 750°F steam.

extra heat exchange surface could be justified up to rather small temperature differences. Based on the 600-psia steam cost developed earlier and the values in Table 3-2, the prices of the steam would be $5.21/1000 lb for 125-psia steam and $3.99/1000 lb for 35-psia steam.

The impact of process temperature (or temperature approach) on the economics of the switch is shown in Table 3-3. Steam flow rates for a 5-MBtu/h exchanger are 5940 pph (pounds per hour) at 125 psia and 5540 pph at 35 psia. Steam savings for 8000 hr/yr are

$$\text{savings} = [5.21(5940) - 3.99(5540)] \times 10^{-3} \times 8000 \approx \$71{,}000/\text{yr.}$$

A reasonable return is possible down to a LMTD of about 20°F, which indicates a good practical set of values.

Enthalpy-based steam values do not provide these incentives. At 125 psia the price would have been $6.86/1000 lb and at 35 psia it would have been $6.72/1000 lb. By inspection, very little investment to reduce steam pressure used in a heat exchanger can be justified with this pricing system.

E. Waste Heat Boilers

Another objective of a good steam price structure would be attractive returns on projects to recover waste heat by generating steam. Other waste heat recovery projects that use the heat directly for process purposes, such as preheating air or reboiling a tower, are generally more attractive than steam generation because capital costs are lower. As a result, recovery of energy as steam is likely to be the most difficult energy conservation option to justify.

If the values calculated for 135-psia and 35-psia steam provide a reasonable return on estimated waste-heat-boiler capital costs, this objective is also met by the availability pricing scheme. A 1977 estimate[2] for a 44,000-pph medium-pressure waste heat boiler on a large flue-gas duct at 450°F was $3.7 million. Escalation to 1983 by the Chemical Engineering Magazine Plant Cost Index would bring the cost to about $5.5 million. The value of the steam generated would be

$$\text{steam value} = 44(\$5.21)(8000) = \$1.83 \text{ million/yr.}$$

From the earlier figures, maintenance would be about 2% of capital and water and chemicals about $0.5/1000 lb, so the net savings would be

$$\text{savings (net)} = \$1.83 \text{ million} - \$0.5(44)(8000) - 0.02(\$5.5 \text{ million})$$

$$= \$1.54 \text{ million/yr.}$$

Simple return is about 28%, therefore, and only marginally attractive.

At lower steam pressures, both capital costs and steam values would be lower. At 35 psia, the steam value would be 75% of that at 125 psia and boiler costs are likely to be only 20% lower. Therefore, it would be harder to justify using the lower-pressure waste-heat boiler. Since part of the purpose of the steam value system is to allow plants to obtain the most efficient configuration possible, this is a consistent result.

F. Steam Pricing Conclusions

Availability pricing of steam represents a thermodynamically sound approach and provides adequate incentives for typical energy conservation projects. In some cases, purchased power costs may be low enough to warrant modifying the steam pricing system somewhat so that efficient back-pressure turbines can be justified.

G. Plants in Which Most Power Is Purchased

For plants in which the vast majority of power is purchased from a utility, the purchased power cost can be used to set the prices for steam at pressures below generation pressure. The value of the lower-pressure steam is calculated according to the following logic: the value equals the cost of steam at generation pressure minus the value of the power which could have been generated in an efficient back-pressure turbine between the generation pressure and the pressure in question.

To demonstrate the method, consider an example in which steam is generated at 600 psig (750°F) and the desired intermediate steam pressure is 140 psia (400°F). In this case the cost of purchased power will be assumed to be 4.4¢/kW·h.

The available energy of the high-pressure steam is 511.9 Btu/lb and that of the outlet 351.3 Btu/lb. Therefore, the work obtained from each pound of steam (ignoring machinery friction losses) is 160.6 Btu/lb. For 1000 lb the value of the power generated is

$$V_p = \frac{(1000)(160.6)(0.044)}{(3413)} = \$2.07/1000 \text{ lb.}$$

Using the value of 600-psig, 750°F steam developed on p. 43, the 140-psia-steam value is calculated by subtracting the value of the power generated:

$$\alpha_{140} = 7.91 - 2.07 = \$5.84/1000 \text{ lb.}$$

The calculated value for 140-psia steam is generally in line with values calculated at other pressure levels using the available energy system. The implication of this is that the price of purchased power is relatively in line with the plant cost for fuel and a conventional condensing power generation heat rate, about 10,000 Btu/kW. In some instances utility fuel is significantly cheaper than plant fuel because it is based on coal or long-term, low-price contracts for hydrocarbon fuels. In these circumstances the energy conservation engineer needs to be wary of basing his utility price system on the price of cheap purchased power, because this price approaches that for the enthalpy value system. He must test the results of the price system to see whether the required incentives for energy conservation investments are still appropriate.

H. When All Utilities Are Purchased

For small plants the entire utility requirement is often purchased from a neighboring plant or utility company. It is in the interest of the seller of utilities to price these products in line with their available energy content. However, the charges for specific plant situations can lead to a skewed price system. It is recommended that prices for purchased utilities be evaluated for consistency according to the available energy ratios or a specifically adjusted formula for the consuming plant which develops recommended utility prices that foster the energy efficiency of the consumer and minimize purchased-utility costs. If the consumer considers himself a part of the utility system of the seller, the analytical procedures developed for the integrated site should apply to him. At the very least, these calculations should provide a basis for negotiation between the two companies to arrive at a mutually beneficial price structure based on their integrated steam power systems.

III. Pricing Other Utilities

A. Refrigeration Utilities Priced on an Available Energy Basis

Most major plant refrigeration systems are mechanical, i.e., driven by compressors. For temperatures in the range of 40°F down to about −70°F, a propane or propylene refrigeration system is generally used. Between −70 and −170°F an ethylene system is cascaded with the propane (C_3) system. This means that the propylene refrigerant is used to condense the ethylene refrigerant and the total horsepower for both refrigeration systems must be calculated for any duty below −80°F. In the liquid natural gas business, there is often a methane refrigeration system cascaded below the ethylene circuit, creating a three-tiered refrigeration system. At the high end of the temperature range discussed, absorption refrigeration systems are sometimes used. These systems consume waste heat and require larger capital investments than the mechanical systems. The value of the refrigeration produced is the same as for the mechanical systems, but the cost elements are significantly different.

Figure 3-5 shows the horsepower requirements for 1.0 MBtu/h of refrigeration as a function of the temperature required in the refrigerant system.[3]

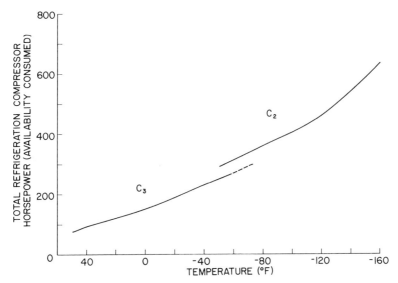

Fig. 3-5 Typical machine power required to transfer 1 MBtu/h of refrigeration at a specified temperature.

Table 3-4

Approximate Mechanical Refrigeration Power Requirements[a]

Refrigerant temperature (°F)	Power requirement (kW/MBtu·h)	Ratio of power required to power required at 40°F
Propylene only		
40	71.6	1.0
0	112.6	1.573
−40	171.6	2.397
−70	223.1	3.116
Ethylene and propylene		
−80	268.6	3.751
−120	358.1	5.001
−160	473.7	6.616

[a] Based on the major propylene–ethylene cascade system with C_3 condensing at 105°F.

A two-stage propylene–ethylene system is shown. The curve for the ethylene (C_2) refrigeration includes the horsepower in the propylene refrigeration system necessary to condense the ethylene refrigerant. The curves represent the typical horsepower requirements for large systems on the Gulf coast of the United States. In northern climates horsepower requirements will be lower because the propylene refrigerant can be condensed at lower pressures. Also, the number of intermediate stages of refrigerant in the system will have an impact on the horsepower costs. Thus, the curves are not meant to provide absolute values of refrigeration costs but rather to illustrate the principle that refrigeration duties can be directly related to power (available energy) consumption.

Power requirements as a function of temperature are shown in Table 3-4. Also shown are the ratios of the power required at other temperatures to the power required for 40°F refrigeration. Following the analysis for the hot end of the system, once the price of incremental power is determined the price at any other temperature level can be calculated in a straightforward manner.

For example, assume power costs 5¢/kW·h. The cost of one MBtu/h of 40°F refrigeration would then be calculated as

$$\text{cost} = 71.6 \text{ (from Table 3-4)} \times \$0.05 = \$3.58/\text{MBtu}.$$

At a temperature of −70°F, the propylene system tabulated in Table 3-4 would require 3.116 times as much power for a cost of $11.16/MBtu. For lower temperatures, both the C_2 and C_3 systems are required and costs are higher still. There is a step change in the curves because of the extra

Table 3-5

Work Equivalents of Heat[a]

Temperature (°R)	W/Q
600	0.105
700	0.233
800	0.329
900	0.403
1000	0.463
1100	0.512
1200	0.553
1300	0.587
1400	0.616
2400	0.776

[a] $T_0 = 537°R$.

irreversibility associated with condensing the C_2 refrigerant with the C_3 stream and then having to recondense that heat load plus heat from system inefficiencies with water in the C_3 condenser. At $-120°F$, 1 MBtu of refrigeration would cost $3.58 \times 5.8 = \$17.9$. Refrigeration is a very expensive utility. In the author's view these relative values should apply to refrigeration produced by absorption processes so that the investment required to convert waste heat into a more valuable utility can be justified.

B. Heat-Transfer-Loop Prices

In general, the value of heat transferred to a process from a heat transfer loop (or hot belt) depends on temperature. Very simply, the available energy (work) content of a parcel of heat at temperature T can be calculated from the Carnot relationship as

$$W = Q[(T - T_0)/T]. \tag{3-2}$$

If $T_0 = 77°F$ (537°R), each million Btu/h of heat can be related to its work equivalent by

$$W/Q = 1 - 537/T. \tag{3-3}$$

Some sample values are calculated in Table 3-5.

Using these ratios the value of any parcel of heat can be calculated from the value of purchased power as before. For example, if heat is available at

$340°$ ($800°R$) and power costs $4.4¢/kW \cdot h$, the price of the heat would be

$$C_H = 0.329(4.4) = 1.45¢/kW \cdot h.$$

C. Value versus Price — The Replacement Approach

So far we have discussed a thermodynamically sound pricing structure for various utilities. This is related to the particular utility stream's capacity to cause desired change (its available energy). In some cases the *value* of a particular utility may be significantly different than its price because it *replaces* a more (or less) valuable utility. This is the economic force that makes for attractive investments. For example, steam can be used to preheat ambient combustion air before it enters a furnace. The net result is that for each Btu of steam heat supplied, 1 Btu of fuel is no longer needed. The value of the fuel replaced is generally greater than that of the steam used. This leads us to an important guideline for identifying economic improvements and projects: *seek ways of substituting utilities of lower available energy content for equal amounts of more valuable utilities.*

A few other examples may help to round out an understanding of the concept. Absorption refrigeration uses waste heat instead of mechanical compression (work) to produce cold. The difference in value between the waste heat stream and the electric power that it replaces may justify the incremental investment of an absorption-refrigeration system. The current difficulty with this process is that much larger quantities of waste heat than power are required, so that the difference in the total available energy input multiplied by its price is much smaller than the price differential alone.

Bottoming cycles, wherein waste heat is converted to power with an organic Rankine cycle, suffer from the same problem. The cycle efficiency is much lower than that for a typical utility plant. Thus, more waste heat and more investment are required per kilowatt. The difference in value for the total amount of waste heat supplied compared with the fuel needed in a typical plant may still justify this investment.

In a number of cases, hot water can substitute for steam in process or space heaters. Again, a lower-priced utility is being upgraded to a use normally served by a more expensive (more capable) source of energy. Here, the one-to-one tradeoff on enthalpy is preserved and the only debit is the amount of heat exchange surface. As we shall discuss later, compensating savings in the utility system may reduce even this debit.

Of course, many designs include applications which *degrade* higher-valued utilities by using them for services when cheaper ones would suffice. This is generally rationalized by investment or reliability considerations which may or may not be valid. More detailed analysis is usually necessary before historical solutions are accepted.

Summary

This chapter showed that the pricing of utilities on an enthalpy basis is theoretically unsound and leads to distortion in the investment practices of the utility system. The available energy content of utility streams satisfies the theoretical requirements for a sound pricing system but must sometimes be modified to distribute the incentives for energy conservation investment equitably and to result in the lowest overall energy cost for a given plant.

Any pricing system should foster energy system improvements. The following three criteria should be met.

(1)   Efficient back-pressure turbine drives are justified.

(2)   The extra heat exchange surface for using the lowest possible steam pressure is justified.

(3)   Recovery of waste heat as low-pressure steam is justified.

For systems in which most of the power is obtained by purchased electricity, the cost of power can be used to set up internal pricing systems for other utilities, again based on the available energy content of the various utility streams. Care must be taken to modify that procedure when purchased-power costs are abnormally low relative to plant fuel costs.

For plants in which the entire range of utilities are purchased from an outside supplier, analysis on an integrated system basis is still pertinent for negotiating prices for the various utilities.

Pricing for refrigeration and other utility systems can also be related to the available energy content of the utility, as measured by the horsepower required to produce it. The power requirements for refrigerants of various temperature and the work equivalents of heat provide ratios that can be used to price these utilities relative to power costs.

The concept that economic projects will result if a less valuable (lower available energy content) utility can be substituted for an equal quantity of a more valuable utility is advanced.

Notes to Chapter 3

1. R. Gaggioli, and W. Wepfer, Available Energy Accounting — A Cogeneration Case Study. Unpublished paper, Marquette University, Milwaukee, Wisconsin, 1978.

2. W. F. Kenney, "Waste Heat Recovery." pp. 61–62. MIT Process Series, MIT Press, Cambridge, Massachusetts, 1982.

3. The data is taken from an internal Exxon Chemical report, but comparable data have been presented in the literature; e.g., R. Mehra, *Chemical Engineering* **85**, 97ff (1978).

4

Characterizing Energy Use

Introduction

The term "energy audit" is loosely applied to a number of approaches to characterizing the energy use in a given plant. With government support there are many seminars given by consultants and engineering companies more or less experienced in performing energy audits. Many of these seminars are very general in nature or aimed at building and commercial energy use. Others concentrate on utility systems and boiler and steam trap efficiencies.

In an effort to appeal to as wide an audience as possible, many purveyors of energy audit technology operate from a collection of general guidelines. In many cases the constraints outlined by the client are accepted without question. One example that underlines the importance of questioning all constraints involved a sound attenuation system which accounted for about 10% of the horsepower consumed in a ventilation system fan. It was included in the design because the owner specified that a sound attenuation system must be applied. An intelligent energy auditor found that the system met all noise criteria without the sound attenuation equipment. Removing the equipment saved 10% of the fan energy with no penalty. Had the design not been reevaluated, the system would have run on forever wasting 10% of its energy input.

I. Understanding Energy Use

To characterize fully the energy use of any plant, we must understand both the needs of the system and the reasons for the needs. The basis for the data gathering must be well thought out. In some locations both summer and winter data are required. For processes that spend a significant amount of time operating away from their design conditions (batch processes, for example), several cases must be considered. Although institutional problems might arise, possible interactions with neighboring facilities may well be profitable to consider.

A. Data Needs

The following list describes the data needs for a comprehensive energy audit:

(1)   a summary, by type, of total utility consumptions;

(2)   unit flow plans (not just schematic, but showing the numbers of heat exchanger shells, process lines, critical instruments, etc.);

(3)   heat and material balances (as up-to-date as time permits) for each unit;

(4)   temperatures and pressures in key parts of the unit [may be included as part of item (1) or (2)];

(5)   separate lists of steam-heated exchangers, furnaces, water-cooled exchangers, air-cooled exchangers, etc., which show heat duties and both designed and "actual" inlet and outlet temperatures ("actual" could be either test data or representative spot data);

(6)   unit plot plan, including adjacent areas;

(7)   utility characteristics (pressure, temperature, amount of sulfur in fuel gas, etc.);

(8)   Utility values (general interest or book) at present and predicted for about the next 10 years (for major, energy-intensive utilities such as steam, fuel gas, etc.),

(9)   steam balance, including any normal steam venting or condensation against cooling water, and condensate recovery;

(10)   product specification considerations;

(11)   any special environmental considerations;

(12)   potential modifications to unit already under consideration, with schedule and basis;

(13)   available spare utility capacities and critical utility limitations (e.g., boiler feedwater, cooling water, etc.);

(14)   list of potential consumers of energy in the vicinity of the unit, if any, including type and level of energy.

For each set of conditions to be considered, the energy input and energy rejection steps must be tabulated along with utility characteristics. This includes internal process heat exchange and chemical heat effects. The flow

Table 4-1

Typical Data List for a Furnace Energy Audit

	Furnace	
Parameter	F-400 Reboiler	F-460 Reboiler
Operating data		
Fuel gas consumption rate (Std. ft³/h)	82,000	34,100
Firing rate (MBtu/h)	59.6	36.3
In the stack		
Temperature (°F)	735	640
Oxygen percentage	3.7	4.0
Flue loss (%)	23.7	22.1
Furnace feed (gal/min)	1386	1217
Temperature (°F)		
Input	370	200
Output	550	320
Vapor percentage		
Input	—	—
Output	—	—
Design firing rate (MBtu/h)	90	36

plan must include adequate pressure and temperature data and the tag numbers of the key instruments that will provide input. Generally, each energy effect is listed separately by type along with the characteristics of the utility consumed or the medium to which heat is exchanged. This is generally accomplished by developing separate lists for each type of equipment, i.e., steam-heated exchangers, furnaces, compressors (and their drivers), etc. Typical examples of such lists are given in Tables 4-1 through 4-5, including data for some hypothetical equipment. The plot plan for the units is an important part of the data. This should include the entire site and all processing and utility units.

Utility system balances and characteristics are also required. This includes things like steam pressures and temperatures, fuel quality and sulfur content, hot-oil temperatures and flow rates, cooling-water flow rates and temperature limitations, etc. A complete steam balance that includes venting, condensation, letdown, condensate recovery, spare turbine or motor drivers, and constraints on turbine operations is very important. The available spare utility capacities and critical utility limitations may also be key parameters.

The values placed upon the various utilities in the plant are very important. As discussed earlier, the various utilities are not always valued correctly from a thermodynamic or energy efficiency point of view. Historical values

Table 4-2

Typical Data List for a Cooling Water Exchanger Energy Audit

| | | Process fluid data | | | | | | Heat duty | | Design data | |
| | | Operating | | | Design | | | | | | Overall heat transfer coefficient U |
Equipment designation	Cooling medium[a]	Flow (lb/h)	Temperature (°F) Input	Output	Flow (lb/h)	Temperature (°F) Input	Output	Operating (kBtu/h)	Design (kBtu/h)	Area (ft²)	(Btu/h·ft²·°F)
E202	CT Water	72,970	143	100	92,642	170	100	1,950	3,954	1,980	91.2
E203	CT Water	83,879	145	116	81,050	130	100	997	998	1,032	78
E205	CT Water	254,972	240	145	233,000	230	130	66,124	63,850	11,300	91
E206	CT Water	94,294	215	145	93,430	220	130	19,733	26,820	4,260	108
E332	CT Water	7,150	245	160	10,100	228	100	7,114	11,000	1,290	89
E335	CT Water	5,497	300	125	5,030	245	130	2,131	1,870	882	37
E362	CT Water	186,679	220	140	196,000	220	160	36,738	38,000	4,450	105

[a] CT stands for cooling tower.

Table 4-3

Typical Data List for a Steam Heat Exchanger Energy Audit

Equipment designation	Steam data				Process fluid data							Heat duty		Design exchanger data	
	Operating		Design		Operating			Design							
	Pressure (psig)	Flow (lb/h)	Pressure (psig)	Flow (lb/h)	Flow (lb/d)	Temperature (°F) Input	Output	Flow (lb/d)	Temperature (°F) Input	Output	Percent vapor	Operating (kBtu/h)	Designed (kBtu/h)	Area (ft²)	U (Btu/h·ft²·°F)
E201	110	5,890	110	10,800	14,137	208	265	173,660	180	300	0	5157	9030	1240	91
E261	110	5,565	90	9,100	7,350	135	210	13,480	163	285	0	4872	7860	768	13
E230	110	3,900	110	6,350	540	116	219	10,510	100	281	22	3414	5800	630	95
E331	110	10,000	110	8,234	—	205	290	9,214	287	287	85	8755	7200	1480	85
E333	110	1,840	110	1,160	—	208	108	5,060	250	250	65	1611	1011	540	19
E434	110	1,160	110	1,540	—	315	316	980	245	245	97	1140	1475	586	8

Table 4-4

Typical Data List for Steam Consumption in Turbines

Equipment designation	Service	Inlet pressure (psig)	Estimated outlet pressure (psig)	Operating steam consumption (lb/h)	Design steam consumption (lb/h)	Rated brake horsepower	Comments
P105	Spare	110	0	1,170	22,600	425	Rolled
P106	Spare	110	0	390	11,800	150	Rolled
P250	Spare	110	0	4,900	27,000	735	Rolled
P162	Spare	110	0	390	4,200	75	Rolled
P100	Spare	110	0	390	5,400	3600	Rolled
P206	Spare	110	0	300	2,350	40	Rolled
C1012	Refrigeration	1600	2	50,000	50,000–165,000	—	—
			125	240,000	50,000–300,000	—	—

Table 4-5

Typical Motor List for Motor Power Consumption

Parameter	Data
Equipment driven	C-201
Voltage	2080
Speed (rpm)	3600
Operating power efficiency (%)	97
Design power (kW)	3000
Designed process control ΔP (psi)	550
Operating ΔP	—
Net process power	—

are often held in high esteem by accountants and may be difficult to change. In addition, forecasts for future fuel prices are often naively done. Yet these forecasts are very important in determining how much investment can be justified by energy efficiency improvement. It is often necessary to go outside the company to the utility companies, fuel suppliers, railroad companies (which transport coal), or even industrial neighbors to develop reasonable forecasts for the next decade.

A further point is the need to reevaluate the product quality specifications imposed on the operations. In the days of cheap energy, "overkill" was a common way of ensuring that products never deviated from specification. At that time the cost was very small, but in these days "overkill," "product give away," etc., can prove to be exceedingly expensive. For cases in which product quality or intermediate product quality is not measured directly but rather inferred from temperature measurements, the acceptability of the quality can be judged only by the frequency of customer complaints. Thus, full specification analyses of typical products and rundowns to intermediate storage are important parts of an energy audit.

B. Precision Requirements

At this stage the required precision of the data is not great. Approximate balances are often more than adequate for a start. Later it may be necessary to go back and develop more precise data in some areas in response to specific ideas for improving energy efficiency. It is also important not to be fooled by calculations to the fourth decimal place supplied by a process computer. In general, balances which come within 10% are suitable for first evaluations. Unaccounted losses in steam systems often run higher than this. In these days such losses should already be the subject of an independent study, but if not, the first item on the energy efficiency improvement list has already been defined.

Note that at this stage we have said nothing about entropy balances around the system. The characteristics of each utility consumed are enough for the moment. Later, from a consistent heat and material balance the required entropy data will be developed, when appropriate.

C. Survey Manpower Requirements

On average, the necessary data for each process unit can be developed in 2 to 4 weeks. New plants are at the low end, whereas old plants with out-of-date instrumentation may exceed the range listed. The personnel required include the contact engineer on each unit, the plant accounting department, some input from the operating supervisor, and generally some energy conservation engineers to work with the data generated. In some organizations a man from the process research and development group would also be an asset.

When possible it is best to supply this manpower from within the company. For a very small plant, 1 to 2 man-months of labor should be adequate. For a major petrochemical complex or refinery, 1 to 2 man-years is the approximate order of magnitude of labor required. Given an effective level of plant involvement, some of the manpower can come from outside if not available within. Contractors and consultants provide this service, but their full role in your program should be thought through before resorting to people not knowledgeable about your plant. In general, lack of commitment by your plant people ("not invented here") will need to be dealt with. In one energy audit led by the engineering department and staffed by consultants, a full energy audit was carried out on a unit that was scheduled to be shut down 2 months after the audit was completed. The operating people were aware of the impending shutdown but neglected to communicate this knowledge even to their own engineering department. The result was to impugn the credibility of the entire effort.

The other danger in using outside personnel is lack of follow-through. There exists a sensitivity on the part of the plant supervisors to any energy survey. In spite of protestations to the contrary, there is always a residual suspicion that the local supervisors will be made to look bad if the potential for significant improvements turns up. This must be dealt with up front and by repeated briefings and assurances to the operating organization. If the operating people fail to have a real ownership interest in the product of the survey, potential improvements will gather dust.

II. Missing Data

Plant measurements are the best source of data, but they are rarely complete. This is particularly true in older units and in utility systems. Before starting the data gathering process, the instrument department needs to recheck the critical meters that do exist to ensure that they are functioning properly. This will never be 100% successful, but to do otherwise invites a considerable waste of manpower. In most cases both the original design and the current operating data are of interest, along with the known reasons for differences. In many cases the extent of fouling is known by the unit engineers and can add considerably to the evaluation of the current status. Major consumptions and other critical inputs must be verified. In some cases the design data may still be close to the truth. For small consumptions the design data may be the only data available. This can often be updated by comparing process temperatures with the design temperatures and performing simple corrections. In this first pass it is necessary to separate the important from the trivial so that an inordinant amount of manpower is not consumed gathering minutiae that will never be used. Many plants set a limit on the size of heat exchangers they will list, e.g., anything less than 1 MBtu/h may be ignored because experience has shown that improving such small items is not economic.

In many cases the use of stoichiometry can fill in the gaps. Consider the tower shown in Fig. 4-1. As for many plants, cooling water rates in this case are not measured and the overhead product comes off on level control. However, since feed rate and composition and overhead product composition are known, much of the missing data can be inferred by calculation.

In this problem p-xylene is to be recovered from a stream containing heavier aromatics. Neither product rate is measured, but feed and reflux rates and the p-xylene content of the overhead are. No heat exchanger duties are measured. With some data from a readily available source, the energy consumption of the tower can be approximated.

From Perry's "Chemical Engineer's Handbook,"[1] we find the following data for p-xylene:

normal boiling point = 138.5°C = 281°F,

latent heat of vaporization = 81.2 cal/g = 146.2 Btu/lb,

and the specific heat = 0.38 at 0°C, 0.43 at 41°C, and 0.55 extrapolating to 140°C.

For heavier aromatics the specific heats range from 0.4 for naphthalene at 87°C to 0.5 for pentadecane at 50°C. Linear extrapolation to 230°C gives specific heats of about 0.8 to 0.9. The product rates can be determined from stoichiometry and one reasonable assumption. The assumption is that

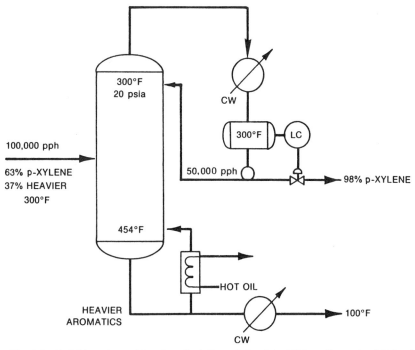

Fig. 4-1 Stoichiometry problem for calculating missing data. CW, cooling water; LC, level controller.

essentially all of the p-xylene exits overhead. Since the bottoms temperature is so high, this must be close to the truth. From component balances the overhead rate can be calculated as follows:

$$0.63 \ F = 63,000 = 0.98 \ M_0,$$

$$M_0 = 64,286 \text{ pph},$$

and therefore

$$M_{\text{bot}} = 35,714 \text{ pph}.$$

The heat rejected in the bottoms cooler will be about

$$Q_{\text{bot}} = 0.8 \times 35,714(454 - 100) = 10 \text{ MBtu/h}.$$

If the heat capacity data were in error, the calculated duty would vary proportionally, i.e., ± 0.1 Btu/lb°F would correspond to about 1.2 MBtu/h. This is likely to be within the precision of other data.

The overhead duty is calculated from the flow rates and the latent heat. Since the reflux is measured, the total flow is 114,300 pph. Therefore, assuming no subcooling, the condenser duty can be calculated as follows:

$$Q_c = 114,300(146.2) = 16.7 \text{ MBtu/h.}$$

If 50°F subcooling occurred, this would add $50 \times 0.55 \times 114,330 = 3$ MBtu/h to the total. This would be important to know. Thus, the reflux temperature should be made part of the data.

The reboiler duty Q_r is the sum of the condenser duty and the heat used in raising the bottoms from 300°F to 454°F:

$$Q_r = 16,700,000 + 35,700(0.8)(454 - 300)$$

$$= 16,700,000 + 4,398,000 = 21,100,000 \text{ Btu/h.}$$

The reboiler duty might range as high as 24 MBtu/h if 50°F subcooling of the overhead occurs. Clearly, subcooling wastes energy unless the external reflux rate is reduced to keep the internal reflux rate constant.

With these approximations, the duties on the cooling water and hot-oil circuits are established. If cooling water (CW) return temperatures can be approximated, the flow rates can be calculated from

$$M_{CW} = Q/\Delta T_{CW}.$$

This assumes that the CW inlet temperature is measured somewhere.

In many cases the accountants and/or utility invoices can supply considerable information. When critical data still needs to be measured, safe and effective procedures must be developed. Infrequent sampling often causes problems with safety and/or validity because of the lack of well-practiced procedures. In addition, techniques used in unit operations laboratories, for example, weighing the condensate from the heat exchanger, may well prove to be unsafe in a plant environment. Such problems need to be thought through in advance and the proper instrumentation provided prior to the time of the audit.

III. An Illustrative Onsite Audit

To illustrate the procedure further, we shall work through developing the data required for the unit shown in Fig. 4-2. From conventional data sources, a petroleum barrel equals 42 gal and the specific gravity of benzene is about 0.8 in the applicable temperature range. Most of the data needed are available on the flow diagrams, but the following heat capacities from the literature[1] will be useful:

$$C_p(\text{benzene at 65°C}) = 0.48 \text{ Btu/lb°F},$$

$$C_p(\text{toluene at 85°C}) = 0.53 \text{ Btu/lb°F}.$$

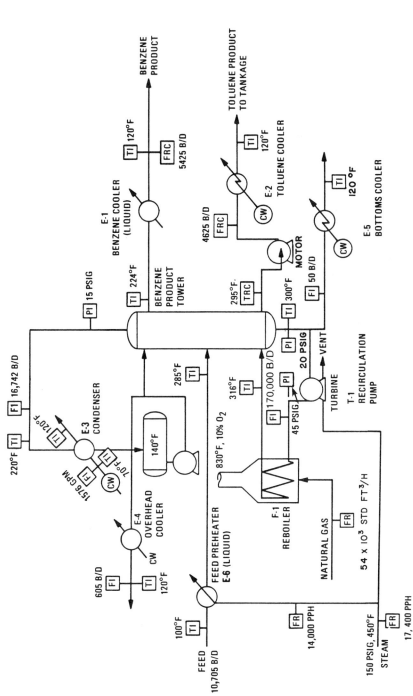

Fig. 4-2 Energy characterization flow diagram for an aromatics product fractionator. CW, cooling water; FI, flow indicator; PI, pressure indicator; TI, temperature indicator; FR, flow recorder; FRC, flow recorder controller; TRC, temperature recorder controller; B/D, barrels (42 gal.) per day.

This process involves the separation of benzene and toluene from a mixed aromatics feed of 10,705 B/d. Energy input comes from a furnace reboiler, a steam-heated feed preheater, and several pumps, one of which is driven by a steam turbine. All surplus heat is rejected to cooling water. Our problem is to characterize the energy use in this unit in such a way as to foster the generation of ideas to improve it. In Chapter 6 we shall actually go through the "brainstorming" process on this unit with the data developed here.

A. Calculation of Heat Balances

E-6, Feed Preheater

$$\text{flow rate} = \frac{10705 \text{ B}}{d} \times \frac{42 \text{ gal}}{B} \times 0.8 \times \frac{8.34 \text{ lb}}{\text{gal}} \times \frac{1 \text{ d}}{24 \text{ hr}} = 125,991 \text{ pph.}$$

From the steam side $Q = 14,000 \times \Delta H_{150} = 14,000 \times 858 = 12 \text{ MBtu/h}$ if the condensate is not subcooled.

E-3, Overhead Condenser

The data for the condenser can be calculated from both the water side and the process side. This provides a valuable check on a major heat rejection step. From the water side, $Q = MC \, \Delta T$, or

$$Q = 1576 \times 60 \times 8.34(120 - 70) = 39.4 \text{ MBtu/h.}$$

From the process side we can assume the overhead is essentially benzene. From the usual source,[1] the boiling point of benzene at 15 psig is 219°F and its latent heat is 162 Btu/lb.

overhead flow = $(16,742/10,705)126,000 = 197,000$ pph,

$$Q = \text{latent heat} + \text{subcooling from boiling point}$$
$$= 197,000 \times \Delta H_v + 197,000 \times C_p(\text{nbP} - 140)$$
$$= 197,000[162 + 0.48(219 - 140)] = 197,000(162 + 38)$$
$$= 39.4 \text{ MBtu/h.}$$

Note that most of the heat is transferred at 219°F. In Fig. 4-3 the temperature profile is plotted against the heat transferred (a $T-Q$ curve). This is second generation data, which is useful in understanding how best to match sources and sinks when beginning the idea generation process.

E-1, Benzene Cooler

$$\text{flow rate} = 5425 \times 42 \times 0.8(8.34/24) = 63,342 \text{ pph,}$$
$$Q = 63,342 \times 0.43(224 - 100) = 3.2 \text{ MBtu/h.}$$

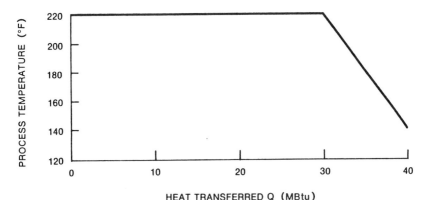

Fig. 4-3 $T-Q$ curve for a condenser.

E-2, Toluene Cooler

$$\text{flow rate} = (4625/5425)(63,342) = 54,000 \text{ pph},$$

$$Q = 54,000 \times 0.53(295 - 120) = 5 \text{ MBtu/h}.$$

The duties for E-4 and E-5 can be calculated similarly, and they are listed in Table 4-7.

F-1, Furnace Reboiler

The amount of fuel fired in the furnace is measured by a meter. The heating value of natural gas is about 1000 Btu/ft^3, so $Q_F = 52,000 \times 1000 = 52$ MBtu/h. The efficiency of the furnace must be calculated, however. This cannot be done from the process side because the extent of vaporization is unknown. The efficiency can be estimated from typical available heat curves for natural gas, which plot stack temperature and oxygen content against either the heat transferred to the process or the furnace efficiency. A sample of such a curve will be shown in Fig. 5-4. Roughly, the O_2 concentration in the stack can be converted to the percentage of excess air by multiplying by five, i.e., 50% in this case. The furnace efficiency is then 73% (where a vertical at 800°F intersects the 50% excess air curve). Thus, the process duty is 0.73(54,000) = 39.5 MBtu/h and stack losses are about 15 MBtu/h.

If an available heat curve cannot be obtained for your fuel, do not despair. Stoichiometry will again come to the rescue. As will be described in Chapter 7, the heat released in the combustion reaction can be calculated from the heat of reaction. The stack losses can then be calculated from the flue-gas flow rate and temperature. For 50% excess air and 1 mol of CH_4 fuel, we

Table 4-6

Overall Heat Balance Check

Component	Duty (MBtu/h)
Inputs	
Furnace fuel	54.0
E-6	12.0
Total	66.0
Outputs	
E-1	3.2
E-2	5.0
E-4	0.07
E-3	39.4
E-5	0.07
Products versus feed	1.1
Subtotal	48.84
Furnace stack (by difference)	17
Total	66

have[1]

$$CH_4 + 3O_2 + 11.3N_2 \rightarrow CO_2 + 2H_2O(g) + 11.3N_2 + 1O_2$$

$$\rightarrow 15.3 \text{ mols flue gas,}$$

$$\Delta H_R = 34,5163 \text{ Btu/lb·mol } CH_4.$$

Discarding the flue gas at 800°F causes the following heat loss to the atmosphere (assuming a heat capacity of 7.5 Btu/mol°F and an ambient temperature of 77°F):

$$Q_L = 15.3 \times 7.5(800 - 77) = 82,964 \text{ Btu/mol,}$$

$$\text{efficiency} = (345,163 - 82,964)/(345,163) = 76\%.$$

This is surprisingly good agreement.

B. Sensible Heat Effects

All products leave the process 20°F hotter than the feed. As a result some heat is carried away from the unit. Roughly, this amounts to 125,000 pph $\times$ 0.45 $\times$ 20 = 1.1 MBtu/h.

The estimate of stack losses can now be checked by an overall heat balance. This is done in Table 4-6. The result is a figure of 17 MBtu/h versus 15 MBtu/h as calculated from the efficiency curve. This is excellent agreement because other heat losses are included in the figure from Table 4-7.

The next step is to tabulate all the data on heat exchange in a form helpful

Table 4-7

Tabulation of Data on Process Heat Exchangers

Component	Duty (net) (MBtu/h)	Flow (pph)	Process temperature (°F)		Percent vapor		Medium	Heat transfer temperature (°F)	
			Input	Output	Input	Output		Input	Output
Inputs									
F-1, Reboiler	39.5		300	316	0	—	Fuel	—	—
E-3, Preheater	12.0	125,000	100	285	0	0	150-psig steam	366	366
Outputs									
E-1, Benzene cooler	3.2		224	120	0	0	CW	70	—
E-2, Toluene cooler	5.0		295	120	0	0	CW	70	—
E-3, Condenser	39.4		220	140	100	0	CW	70	120
E-4, Overhead cooler	0.07		140	120	0	0	CW	70	—
E-5, Bottoms cooler	0.07		300	120	0	0	CW	70	—

Table 4-8

Pump Power Requirements

Driven Equipment	Operating characteristics					Design characteristics		
	Volts	Amperes	Kilowatts	Flow rate	ΔP	Flow	Head	Efficiency
Motors								
Reflux pump	—	—	—	195,000	—	—	—	
Toluene product	—	—	—	54,000	—	—	—	

	Resources		Steam rate	Process		Comments
	Inlet	Exhaust		Flow	ΔP	
Steam turbines						
Reboiler pump	150 psig	Atmosphere	3400 pph	198,500 pph	25 psi	Apparent efficiency is 30%

to those charged with improving the process. A typical arrangement is shown in Table 4-7. The key data necessary to analyze the unit and suggest improvements to heat exchange systems are shown. The furnace is included here for convenience.

C. Pump Power

There remains the task of calculating the power consumed in the pumps. The larger pump is driven by a venting steam turbine. The 150-psig steam flow is derived by difference from the steam meters at 3400 pph. The process horsepower can be approximated from

$$W = \frac{(\text{B/d} \times 42 \times 7 \text{ lb/gal}) \, \Delta P}{(24)(60)} \times 33,000 \times 144 \times \frac{1}{33,000} = 72 \text{ hp.}$$

From standard tables[2] of the theoretical horsepower available from exhausting 3400 pph of 150-psig steam at 450°F to the atmosphere, we find that 18.22 lb of steam should produce 1 kW·h. Thus, the efficiency of the turbine plus the efficiency of the pump is ~30%.

No data are provided with which to calculate the power or efficiency of the other two pumps. Their power consumption must be measured in the field by electricians. No discharge pressure indicators are shown in the flow diagram, which is not unusual. A trip to the field will clearly be necessary to complete Table 4-8, in which pump power requirements are listed.

The data are now in a form for evaluation. We shall continue with this problem in Chapter 6.

IV. An Illustrative Steam Power Balance

Steam power balances are usually shown in the format presented in Fig. 4-4. Major steam flows are generally measured, but losses, letdowns, and deaerator flows often cause problems. Much of this data can be calculated from heat and material balances, but some must be guesstimated or combined into calculable quantities. Calculations for the balance given in Fig. 4-4 will be carried out as a practical example.

Flows shown on the diagram are generally net flows to the consumer. One should first check that all boiler auxiliaries are shown on the diagram. If not they can either be added to complete the balance, or a factor can be determined for gross boiler steam make versus net flow and used to correct boiler fuel requirements. For this case we shall assume that all boiler auxiliaries are included in the diagram.

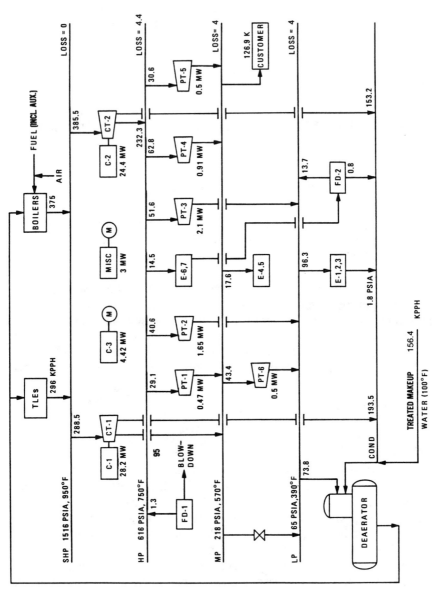

Fig. 4-4 Format for a steam power balance. All flows are in kilopounds per hour.

The first step is to do a mass balance at each pressure level. At 1516 psia the balance is simple and losses are zero:

Input (kpph)	Output (kpph)
TLEs = 296	CT-1 = 288.5
Boilers = 378	CT-2 = 385.5
Total = 674	Total = 674

At 616 psia we can also calculate all quantities and losses are 4.4 kpph:

Input (kpph)	Output (kpph)
CT-2 = 232.3	PT-1 = 29.1
FD-1 = 1.3	PT-2 = 40.6
Total = 233.6	E-6,7 = 14.5
	PT-3 = 51.6
	PT-4 = 62.8
	PT-5 = 30.6
	Total = 229.2

Around the 218-psia level indeterminate quantities start to develop. The balance is as follows and letdown and losses are equal to 29.6 kpph:

Input (kpph)	Output (kpph)
CT-1 = 95	Letdown = ?
PT-1 = 29.1	PT-6 = 43.4
PT-4 = 62.8	E-4,5 = 17.6
PT-5 = 30.6	Customer = 126.9
Total = 217.5	Losses = ?
	Total = 187.9

The next step is to see if the 65-psia header balance will help solve the problem:

Input (kpph)	Output (kpph)
Letdown = ?	Deaerator = 73.8
PT-6 = 43.4	E-1,2,3 = 96.3
PT-2 = 40.6	Losses = ?
PT-3 = 51.6	Total = 173.1 + losses
FD-2 = 13.7	
Total = 149.3 + letdown	

There are still too many unknowns for the flows to be calculated separately. The best approach is to measure the flow through the letdown valve. If this is not possible, an assumption must be made before the balance can be completed. If it is assumed that losses are the same from all pressure levels, the letdown flow can be approximated. From the 218-psia balance,

$$\text{letdown} + 4 = 29.6,$$

$$\text{letdown} = 25.6.$$

From the 65-psia balance,

$$\text{letdown} + 149.3 = 173.1 + 4,$$

$$\text{letdown} = 27.8.$$

This is close enough to give an indication of the power that may be wasted in the letdown.

The condensate balance is next. Condensate is not recovered from the medium pressure customer. The deaerator water balance is as follows and the makeup $= 674 - 517.8 = 156.4$ kpph:

Output (kpph)	Input (kpph)
Boiler feed water = 674	Low-pressure steam = 73.8 CT-1 = 193.5 CT-2 = 153.2 E-1,2,3 = 96.3 FD-2 = 0.8 Makeup = ? Total = 517.8 + makeup

The completed steam power balance is shown in Fig. 4-4. Some 346,700 pph of steam flow from turbines into condensers. This represents more than half of all the high-pressure steam produced.

Summary

Developing approximate unit energy balances and site steam balances is the first step in understanding plant energy use. A quest for excessive precision should be avoided in the early stages until some data analysis identifies the systems for which more precise data are required. Stoichiometric calculations can do much to fill in missing data. Care should be taken to ensure that any extraordinary samplings or equipment tests are carried out safely.

To assist in developing new ideas for saving energy, the audit data are usually arranged in tabular form. This facilitates the matching of sources and sinks and the identification of the larger inefficiencies.

Steam power balances are an important part of the data necessary for a successful brainstorming session. They should be complete enough to show inefficiencies, but some approximations will normally be required. In many cases, steam balances will be needed for both summer and winter.

The use of these data to identify energy saving opportunities will be pursued in Chapter 6.

Notes to Chapter 4

1. J. Perry, ed., "Chemical Engineer's Handbook," 3rd ed. McGraw-Hill, New York, 1950.
2. R. C. Spencer, P. G. Rossettie, and R. B. McClintock, "Theoretical Steam Tables." American Society of Mechanical Engineers, New York, 1969.

5

Optimum Performance of Existing Facilities

Introduction

Most plants do not run day in, day out at the best possible energy efficiency (or catalyst efficiency or product recovery, etc.). This has not only been my observation, but has also been the subject of many articles and meeting presentations. Getting closer to the best possible operation for more of the time shows up immediately in the profit column. In addition, a sound base case is needed against which to evaluate any investment ideas for saving energy. Thus, the first step in reducing plant energy use is to understand the real limitations of the installed hardware, the control systems, and the people and approach them as closely as possible.

The specific reasons for failing to operate at optimum conditions are innumerable. In a gross oversimplification I have classified them into two categories:

(1) awareness, motivation, and diligence;
(2) technical and management information.

The second category can affect the first. It is very hard to be diligent about saving energy if you do not know what to do or where to look. Because this is a technical book and because the psychology of each plant management situation is different, I will deal with nothing more than a few passing comments on the first category. This chapter will discuss some basic technical information on reducing losses of available energy and develop a technical basis for determining optimum energy use and controlling closer to it.

I. Principle 1 — Minimize Waste

This is an obvious "motherhood" statement. However, there are times when the fundamental technical – economic information is not understood *at the level where decisions are made.* Consider what an energy conservation consultant found at a major high-energy physics laboratory. Makeup cooling water was being pumped up to a pond around the main accelerator. Two 1500-gal/min pumps blasted away with an overflow back to the reservoir. The consultant found *on average* that only 750 gal/min were needed to make up for evaporation. Without this knowledge, the operators kept pumping the maximum because that was the way the system had been designed. With the new information, one pump could be shut down until there was no overflow. Thereafter, the pumps could be throttled on the same criteria or some instrumentation could be installed to indicate and/or control the flow needed. In a single pumping installation over $1200/yr were being wasted even though the system was running "at design." The point is that "design" is usually set by the worst set of conditions to be met and more information is required to operate the unit in an optimum fashion.

Another comment about "design" is also worthwhile. Over time systems tend to wander off design in the wrong direction for various reasons. In a 50,000-std. ft³/min ventilating system designed for 6 in. of water pressure drop, good balance could not be obtained. Advantage was taken of excess blower capacity to speed up the fans and increase the head to 10 in., where the system could be balanced. The plant ran that way until an energy audit was carried out that questioned the practice. The energy conservation consultant who did the audit worked the base-case balance problem and reduced the fan head to the design figure. Savings were in the range of $50,000/yr, simply for restoring the system to the "design" ΔP.

Both of the situations described are typical of the kind of things which go on in every plant evey day. Start-up safety margins are never tightened up and/or operating conditions slowly evolve away from optimum little by little. With enough technical data in the right place much of this wasted energy can be saved.

Combustion and steam systems are primary sources of more or less "invisible" waste. The key issues are usually too much excess oxygen and multiple small losses, respectively. Many times the magnitude of these losses are not clear to either the operators or the management.

II. Combustion Principles

The vast majority of simple heat losses in a furnace or boiler go up the stack. To the extent that these losses cause more fuel to be fired, they have a significant effect on available energy consumption. Figure 5-1 shows a generalized diagram of a heater. All incoming air and fuel are heated to

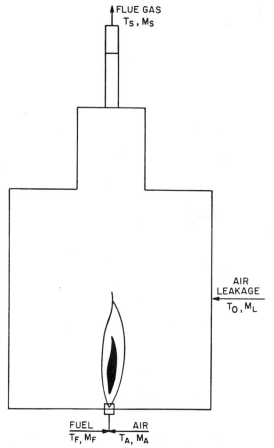

Fig. 5-1 Typical heat effects during heater combustion.

combustion temperature and then cooled to the stack temperature by process needs. Since most furnaces run at a vacuum on the combustion side, additional in-leakage of air occurs, which must also be heated to stack temperature and therefore adds to the loss. Stack heat losses can be calculated as follows:

$$Q_S = M_F C_F (T_S - T_O) + M_A C_A (T_S - T_O) + M_L C_L (T_S - T_O), \quad (5\text{-}1)$$

where M_F, M_A, and M_L are the mass flow rates of fuel, combustion air, and leaks, respectively, T_S and T_O are the temperatures of the stack gas and the ambient (outside) air, respectively, and C_F, C_A, and C_L are the heat capacities of the streams.

In general, an averaged heat capacity is used to calculate heat losses. This is generally accurate enough because the vast majority of the flue gas is nitrogen. The equation thus simplifies to

$$Q_S = C_{mean}(M_F + M_A + M_L)(T_S - T_O) = M_S C_{mean}(T_S - T_O). \quad (5\text{-}2)$$

A. Optimum Excess Air

In the most efficient and ideal combustion of fuel, the fuel is burned completely to CO_2 and H_2O with 0% excess air so that there is no oxygen leftover. With industrial burners, however, if stoichiometric combustion is attempted, carbon monoxide and unburned hydrocarbons are produced. When extra air is used for combustion, energy losses increase for two reasons. First, the extra air requires additional fuel to heat it from ambient to stack temperature. Both changes increase the flow of the flue gas. Second, the extra air cools the flame in the radiant section, reduces the amount of heat transferred there, and lowers the initial available energy of each pound of the flue gas. In a given design with constant process duty, this can only be compensated for in the convection section by having a higher stack temperature. Thus, excess air increases the fuel rate because of the greater amount of hot combustion gas leaving the furnace, and it also raises the temperature of the exhausting combustion gases. Figure 5-2 shows the incentive for reducing excess O_2 to 4% in the stack gas from higher values at a variety of stack temperatures. Savings are calculated at $4/MBtu. As an example, reducing O_2 from 10% to 4% with a 600°F stack temperature saves about $250,000/yr per 100 MBtu fired. A 4% O_2 percentage is a reasonable value for typical industrial liquid fuels. For clean liquids and gases 2% is a better target. Large boilers with sophisticated controls often target 1.5% O_2 or less.

The goal of efficient combustion is to keep the percentage of excess air as low as possible. There is a lower limit, however, that is reached whenever there is incomplete combustion or flame impingement on the tubes. Incomplete combustion is undesirable because not all of the potential heating value of the fuel is used. With the heavier fuels, this can be visually determined by the smoke coming from the stack (smoke being made up of minute carbon particles). With lighter fuels, incomplete combustion can occur without smoke but can be determined by the carbon monoxide (CO) concentration in the stack gas. When fuel burns to CO rather than CO_2, only one quarter of its potential heat is released. A rule of thumb used to gauge the inefficiency caused by incomplete combustion is that for every 1% CO in the flue gas, 3 to 4% of the fuel is wasted. CO concentrations high enough to be detrimental to efficiency usually do not occur above oxygen contents of $\frac{1}{2}$% (2 to 3% excess air) in the combustion products. (See Fig. 5-3.)

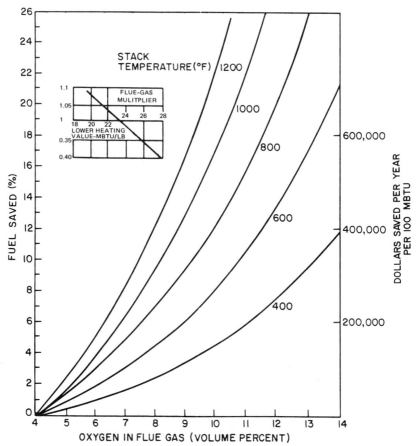

Fig. 5-2   Fuel savings in furnaces obtained by reducing the amount of excess oxygen in the flue gas to 4%. The data are based on oil fuel (for gas a multiplier must be applied) and include the effect of reduced stack temperature (±10%). The service factor is 95%; the fuel cost is $4/MBtu.

Too little air can cause damage to furnace tubes as well as waste energy. Ultimately, the flame shape deteriorates to the point that hard flame impinges on the tube. This can result in local hot spots, damage, and coking. The best way to determine if hard flame impingement is occurring is to view the firebox using blue tinted glasses made specifically for that purpose. These glasses eliminate the glare and bright haze that make it difficult to observe the real position of an oil flame.

Another consequence of reducing excess air too far is afterburning in the convection section. This condition leads to elevated tube metal tempera-

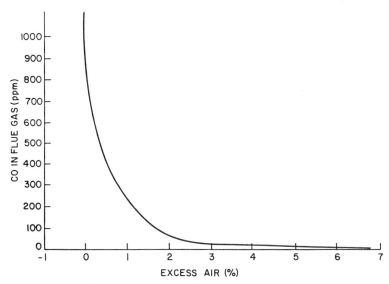

Fig. 5-3   Typical carbon monoxide–excess air relation in stack gas.

tures when unburned hydrocarbons burn with air that leaks through the upper portions of the furnace. Elevated tube metal temperatures in turn lead to accelerated corrosion rates, potential premature tube failures, and the sagging of horizontal tubes. Obviously, these conditions have to be avoided as well as those in which too much excess air is used.

B. Control of Draft in a Furnace or Boiler

Ideally, furnace draft is controlled to maintain a sufficient vacuum in the furnace to draw air in through the burners. The air registers on the burners are adjusted to control the amount of air. In setting draft the objective is to avoid a very high draft that would cause excessive air leakage into the furnace and to avoid a very low draft that would cause positive pressure underneath the convection tubes and cause hot flue gas to leak out of the furnace. However, when the furnace is well sealed and air leakage is not a problem, draft can be used to control the amount of air entering the burners.

Figure 5-4 shows the draft and pressure balances in a typical furnace using natural draft burners with a 200-ft (61-m) stack. Line A on the graph is a plot of the ambient air pressure at this level. For every 200 ft of height, the pressure reduces by about 0.1 psia which is equivalent to a 3-in. or 75-mm water column.

Because the stack is open at the top, the pressure inside is the same as outside, 14.6 psia. At a stack temperature of 500°F, the gas inside the stack

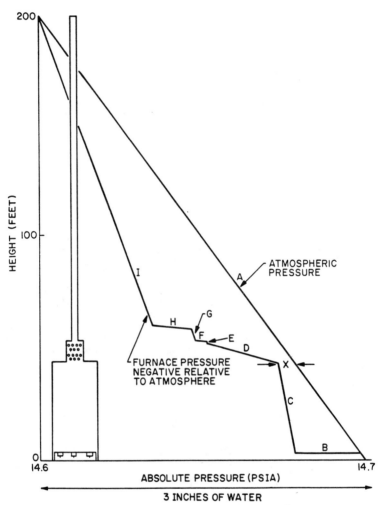

Fig. 5-4 Furnace draft: A, atmospheric pressure versus height; B, pressure drop across the burner; C, height loss; D, pressure drop across the convection tubes; E, height loss; F, pressure loss entering the stack; G, height loss; H, pressure drop across the damper; I, height loss; X, draft at the arch.

has only half the density of the surrounding cool air. Therefore, the pressure inside the stack (line I) increases at only half the rate outside the stack (line A). This creates the draft or vacuum in the furnace.

The pressure balance in the furnace starts with line B, which represents the pressure drop of the combustion air as it enters the burners through the air doors. Line C represents the pressure loss due to the height of the radiant section. This pressure loss is very slight in the radiant section because the gas is very hot and has a low density. Line D is the height loss in the convection section plus the frictional pressure drop as the gas passes through the tubes. Line E is the height loss above the convection tubes just before entering the stack. Line F is the pressure loss entering the stack and equals one-half velocity head for an entrance loss plus one velocity head for converting static pressure into the velocity in the stack. Line G is a height loss, and line H is the pressure drop across the damper.

In Fig. 5-4 the pressure drop across the damper is shown at 0.4 in. If the damper were opened completely, the pressure drop across the burners would increase and line C would become more negative and increase air leakage into the furnace (which is undesirable). Alternately, if the damper were closed more, X, the vacuum under the convection tubes, would decrease. The damper can be closed enough so that X becomes positive and hot flue gas leaks out of the furnace and causes the structural steel to sag and oxidize.

Proper control of furnace draft means the damper should be adjusted so that the pressure X underneath the convection tubes *is a 0.1-in. or 2- to 3-mm water column vacuum.* When strong or gusty winds cause the draft in a furnace to fluctuate, the damper should be opened slightly so that the vacuum is not less than 2 mm during the fluctuations. Also, in a wind the part of the furnace facing into the wind is exposed to the highest static atmospheric pressure. If the draft measurement made at the point of highest atmospheric pressure was normal, the other sides of the furnace, which are exposed to a lower static atmospheric pressure, might have too little draft or perhaps even a positive pressure. In a 17-mph wind the difference between front-face and side-static atmospheric pressure can be 0.1 in. of water. In a 30-mph wind this difference can increase to 0.3 in. of water. Thus, in large furnace multiple draft gauges and a method of selecting the smallest reading may be necessary to prevent positive pressure and resultant damage or safety hazards.

If additional draft gauges are needed, or if one is not working, a simple U-tube manometer connected to the furnace by plastic tubing will suffice. New furnaces often have draft gauge connections above and below the damper. These are not needed for operating the furnace; they are only used for trouble-shooting or for checking draft calculations.

Stack draft is markedly affected by stack temperature. For many of today's efficient furnaces, we find that induced draft fans are more economical than very tall stacks in providing the required draft. Table 5-1 shows the variation of required stack height with flue-gas temperature for one desired total draft. Reducing the flue-gas temperature from 700°F to 300°F doubles

Table 5-1

The Effect of Flue-Gas
Temperature on Required Stack
Height for a 0.8-in. H_2O Draft

Temperature ($°F$)	Required Height[a] (ft)
300	240
400	185
500	160
600	143
700	130

[a] Includes stack friction and exit loss
of 0.2 in. H_2O.

the required stack height. However, most of that increment occurs as the
temperature drops from 400°F to 300°F. The potential for optimization,
particularly in retrofit situations, is obvious.

C. Oxygen Analyzers

Many modern furnaces have both forced-draft fans to control air to the
burners and induced-draft fans to control draft. This often allows oxygen
control by direct measurement of air and fuel flow rates. Large, efficient
process furnaces with natural draft burners often have induced-draft fans
because stack temperatures are too low to get the required draft with a
practical stack height. Control systems are devised to maintain the desired
amount of excess air. All such systems involve oxygen analyzers in the stack.
Sometimes CO analyzers are added to allow even more precise control.
Damper settings are adjusted automatically to control O_2 subject to a
limitation on absolute draft level. In-stack analyzers such as the Westing-
house system to be described later are the most reliable. Please note that
relatively small (10 MBtu/h) furnaces can also justify O_2 analyzers for
energy conservation.

The high reliability, low maintenance, and fast response (less than 1 sec)
of the in-stack analyzer are a great improvement. The design allows the
analysis section of the instrument to be mounted directly in the furnace
flue-gas stream. The oxygen content is sensed by a heated (1500°F) electro-
chemical cell at the tip of a 1½-, 3-, or 6-ft probe. In-stack mounting
completely eliminates the need for a sample handling system and is the
major factor contributing to the superiority of this type of analyzer over
conventional types.

D. Location of the Oxygen Sample Point

Normally, the permanently mounted oxygen analyzer sample point should be located downstream from the convection section to obtain as uniform a reading as possible. This is especially important when the furnace is on automatic control. If the sample location were in the radiant section where the flue gas from different burners is not well mixed, the analyzer reading would mainly reflect the operation of the burners close to it. If any of those adjacent burners were taken out of service for cleaning or adjustment, some error could be made in the oxygen reading. The exceptions to placing the oxygen analyzer downstream from the convection section are furnaces that have more than one service in the radiant section. Here, correct adjustment of excess air in each cell can only be made by having one analyzer in each cell. Also, in furnaces with high tube metal temperatures in the convection section, it is desirable to monitor radiant section oxygen to avoid afterburning. In this case at least three probes are required to achieve a representative analysis of flue gas. The placement of the oxygen analyzer downstream from the convection section does require that there be minimum leakage of air into the convection section to avoid a false high oxygen reading, i.e., furnace maintenance becomes more important.

Some common causes of too much excess air are shown in the following list:

(1) improper draft control;
(2) leakage through openings;
(3) faulty burner operation,
    (a) dirty burners,
    (b) poor adjustment – maintenance on air doors – controls,
    (c) dual fuel burners needed.

Clearly diligence is required as well as the proper controls. Diligence is a function of plant morale and motivation, not technical subjects. However, better morale will often result when the requisite technical information is made available to operating and maintenance people in an easily assimilated form.

III. Illustrative Problems —
Combustion Efficiency

To illustrate the importance of excess air and its interrelation with stack temperature, we shall consider the following problems, using Fig. 5-5 to develop solutions.

Table 5-2

Impact of Excess Air Changes on Fuel Efficiency, Fuel Requirement, and
Fuel Savings[a]

Stack conditions	Efficiency (%)	Fuel requirement (MBtu/h)	Fuel savings (MBtu/h)
1000°F, 200% excess air	47	213	Base
300°F, 200% excess air	87	115	98
1000°F, 20% excess air	77	130	83
300°F, 20% excess air	94	106	107

[a] Data from Fig. 5-5.

(1) Calculate the fuel required to provide 100×10^6 Btu/h to a process if the flue gas is at $1000°F$ and combustion takes place with 200% excess air.

(2) How much energy would be saved if only the stack temperature were reduced to $300°F$?

(3) How much energy would be saved if only the excess air were reduced to 20%?

(4) What would the total energy savings be if excess air could be reduced to 20% in addition to reducing stack temperatures to $300°F$?

The results of these problems are given in Table 5-2. The result for Problem 1 is diagramed in Fig. 5-5. From the graph the process heat efficiency is seen to be 47%. The amount of fuel fired to provide 100 MBtu/h to a process would be

$$Q = 100/0.47 = 213 \text{ MBtu/h}.$$

The other cases can be calculated in the same way.

Obviously, excessive combustion air is more wasteful at high stack temperatures than at low ones. Stack oxygen levels should thus be monitored more closely on old, less efficient furnaces and boilers because the incentives are greater. Investment is required to improve stack temperatures. This may well be justified, but money can be saved *today* by reducing excess oxygen to the minimum. This effort will provide a proper basis for evaluating any new investment as well.

The direct connection between changes in the amount of fuel fired and the available energy expended on a process will be a recurring theme in this book. There is no substitute for understanding the net impact on the fuel fired at the site for any change in operating conditions or facilities. This is the net thermodynamic impact, and generally the only place where energy conservation efforts can be measured directly in dollars. As systems become more complex, this impact becomes more difficult to trace. Several methods of inferring the answer have been suggested so far, and more will be

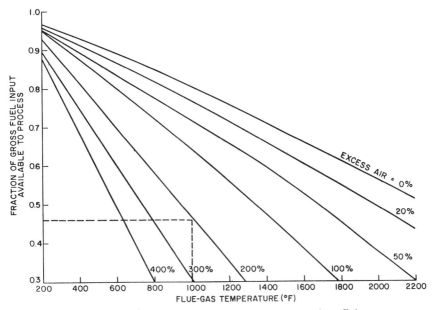

Fig. 5-5 Available heat diagram for calculating combustion efficiency.

discussed in later chapters. For major decisions it is usually advisable to work back to the fuel impact to be sure.

IV. Steam Trap Principles

Steam waste, which causes unnecessary boiler firing, often consists of many small leaks, some to the atmosphere and some across steam traps into closed condensate recovery systems where they are almost truly "invisible." Figure 5-6 shows rates of steam loss for various sizes of leaks at several steam pressures. A $\frac{1}{4}$-in. leak in a 600-psig line loses over 1000 lb/h. At the costs calculated in Chapter 3, this leak would amount to about $6/h, or $50,000/ yr, aside from the associated noise, safety, and physical damage problems. Maintenance priorities should be set on current profit losses, not old rules based on low fuel costs.[1]

Steam traps have been the subject of much discussion and advertising, yet they remain something of a mystery to many people. This book cannot attempt to summarize or evaluate all of the data in the literature. Some fundamental insight into what a steam trap is and how it works, however, will be attempted.

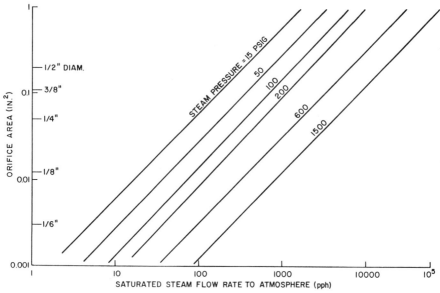

Fig. 5-6 Estimated steam losses at varying leak sizes and steam pressures in a steam trap.

A. The Three Functions of Steam Traps

The first practical attempts to accomplish what a steam trap does involved manual draining of the condensate. This progressed rapidly to use of a throttled drain valve, which cost dearly in steam losses. The steam trap thus fulfills the same role as an automatic control valve that is sensitive to load. The functions of a steam trap are as follows:

(1) to eliminate air and other noncondensable gases from the heat exchange system because they tend to insulate heat transfer services and reduce the rate of heat transfer;

(2) to eliminate condensate from the system, which will also blanket heat transfer services or increase the pressure drop in steam transport lines;

(3) to accomplish the foregoing with the absolute minimum loss of live steam from the system.

In addition, a good steam trap does not freeze in cold climates, lasts a long time and requires minimum maintenance, can be standardized so that a minimum number of spare parts is required, can be maintained in line, and is adaptable to the full range of loads placed upon it in the given application.

B. The Three Basic Types of Steam Traps

There are three basic types of steam traps in practical use: mechanical, thermostatic, and thermodynamic. There are also some combinations on the market, but these are more expensive than simpler models. There are many manufacturers producing similar models with slight variations. A brief summary of the three major types of traps is shown in the following list:

(1) mechanical traps, which operate on differences in density,
    (a) inverted bucket,
    (b) float;
(2) thermostatic traps, which operate on temperature differences,
    (a) bellows,
    (b) bimetallic;
(3) thermodynamic traps, which operate on vapor–liquid changes,
    (a) disk,
    (b) piston.

The principles of operation for the major types and their inherent qualities are described in the following paragraphs.

1. Mechanical Traps

Mechanical traps are operated by the difference in density between steam and condensate. The first models employed a *float* attached to a lever that was connected to the discharge valve. The movement of the float adjusted the position of the valve to regulate continuously the outflow of condensate. Over 40 years ago the Armstrong Machine Works pioneered the development of the inverted bucket trap (see Fig. 5-7). Steam inside the bucket causes it to rise, ultimately closing the effluent valve. Noncondensables (and a little steam) pass through a small hole in the bucket to the discharge.

2. Thermostatic Traps

These traps respond to temperature changes in the line, opening when condensate reaches the thermostatic element and closing when the hotter steam arrives there. Basically, *the trap senses the difference in temperature between live steam and condensate,* which has cooled somewhat owing to the heat losses from the system. Thermostatic traps are often actuated by a fluid inside a bellows mechanism (see Fig. 5-8a) or by the bending of a bimetallic strip or strips (see Fig. 5-8b). Careful design of the bimetallic elements is required so that the force available to close the valve will be adequate to overcome the system pressure.

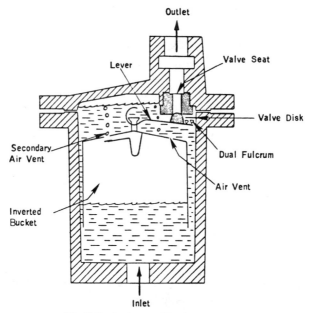

Fig. 5-7   An inverted bucket steam trap.

Thermostatic traps are wide open when cold, so they provide quick startup for the equipment by eliminating air and cold condensate at the full flow rate. *When operating at temperature, however, response is more sluggish because some time is required for the absorption of heat to open and close the valve.*

In many cases thermostatic and float elements are combined in a single trap. The addition of the thermostatic element to a float trap provides much greater air handling capability than is possible with the float mechanism alone. In this way it is possible to take advantage of the variable flow characteristic of mechanical traps with variations of condensate load and still maintain good air handling capacity.

3. Thermodynamic Traps

The third type of trap in common use these days is the thermodynamic trap. *The operating principle here is that as condensate flows to the low-pressure side of the trap, steam is flashed off owing to the pressure drop.* The resultant large increase in volumetric flow causes the ΔP in the outlet section to increase, generating a back pressure of steam tending to close the valve. With *cool* (15–30°F below saturation) condensate in the trap, the valve remains open to allow the passage of water from the system. The same is true

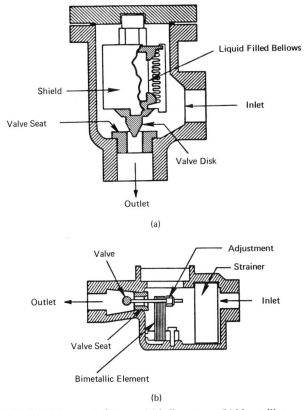

Fig. 5-8 Thermostatic traps: (a) bellows type; (b) bimetallic type.

of air, so that the valve remains fully open on startup until hot condensate reaches the chamber. When this occurs appreciable fractions of the water flash to steam, building up the pressure in the control chamber and closing the trap. A schematic diagram of a disk-type thermodynamic trap is shown in Fig. 5-9. Thermodynamic traps are normally limited to a back pressure no more than 30% of the inlet pressure or the trap could fail to function.

C. Drain Orifices Pass a Fixed Amount of Condensate or Steam

Drain orifices are plates with a small hole, typically 0.51–6.35 mm (0.020–0.250 in.), which pass condensate at a given rate but pass only a limited amount of steam because of the throttling effect of steam passing

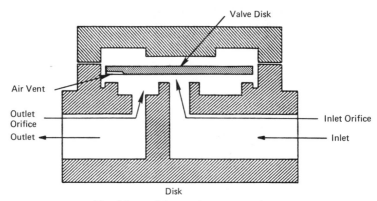

Fig. 5-9   A disk-type thermodynamic trap.

through the orifice at high velocity. Drain orifices waste less steam than any other type of steam trap that discharges condensate at saturation temperature. However, they are sized for only one condensate load. They are, therefore, suitable only for a fixed condensate load such as occurs on a steam main.

D. Steam Trap Sizing Is Not Based on Pipe Size

The proper selection of a steam trap depends on six factors:

(1)   the anticipated condensate loads,
(2)   the pressure differential across the trap,
(3)   start-up load,
(4)   the temperature of the condensate reaching the trap,
(5)   the need to vent air or noncondensable gases,
(6)   the proper safety factor for the application.

You will notice that nothing was said about the size of the piping connections at the inlet and outlet of the trap. It is not unusual for people to feel that trap selection is dictated by pipe size. This can be catastrophic. Oversizing of the steam trap often contributes to early failure through excessive cycles and wear on internal parts. In addition, thermodynamic traps will leak live steam at less than 10% of capacity. Therefore, the proper selection of a steam trap for a given duty requires, first, the definition of the duty, second, a definition of the variation in the conditions of service and, third, analysis of the technical details of the specific equipment being considered. *Do not let a purchasing clerk design your steam system!*

E. The Proper Safety Factor Varies with Application

All manufacturers have what they call a safety factor to be applied to trap sizing. For nonprocess loads this generally involves a factor of two on the anticipated condensate load. It is meant to cover abnormal loads or factors not allowed for in the design of the plant. Also, the safety factor involves margins of error for lower pressures at startup or other times during the working period, the possibility of slugs of water arriving at the trap, or the possibility of increased back pressure caused by leaking steam traps elsewhere in the condensate system. Another point is that most steam traps work on an intermittent basis, whereas the capacities quoted in catalogs are on a continuous basis.

Back pressure has a major effect on all trap operation. Increasing back pressure will reduce capacity up to the limitation of the design. The source of this back pressure is also important. If it results from a steam bypass and/or leaking traps into the condensate return header, the increased temperatures would cause bimetallic traps to close, but disk traps would stay open and blow steam.

Both the thermostatic and thermodynamic traps can be arranged to be freezeproof in normal operation. In addition, some varieties of traps can also be freezeproof while not in operation, because they drain fully and the valve remains open when no steam is flowing. This is an important point in northern installations.

F. Conclusion

From the standpoint of trap selection, there are a number of solutions available for most problems. A key point to remember, however, it that a poorly sized or installed trap or a trap that is not maintained will cost you much more each year than small differences in performance among the various designs, all operating properly. Another key factor to consider is that vendor service and space parts supplies will be essential to a proper maintenance program.

With these qualifications, some suggested selection guidelines are the following.

(1)   Simpler is better and proper sizing is essential.

(2)   Mechanical traps are subject to freezing. Explore the situation with each vendor. They are often the only trap practical for low-pressure services.

(3)   Disk-type thermodynamic traps seem most economical for steam drip legs, tank coils, steam tracers, and other light to medium duty, lower-pressure loads above 20 psig.

(4) For high back-pressure systems, thermostatic traps are often the most appropriate. Bimetallic units offer maintenance advantages and greater flexibility *once the trap is properly adjusted.*

(5) Bimetallic traps also have advantages in high-pressure systems, in dirty systems, in systems with a large noncondensable content, and in systems subject to superheated steam.

(6) Process applications require consultation with the vendor's technical people before selection can be made. Selection by price alone is short-sighted.

(7) A sufficient number of vendors now offer units which can be maintained in line and which utilize standardized spare parts. This type of trap has many advantages for an efficient maintenance program.[1]

General Note: Other forms of energy waste are often related to the inefficiencies in individual unit operations. A number of these operations will be discussed in later chapters.

V. Principle 2 — Manage Energy Use Effectively

Almost every operations and/or energy manager believes he does an excellent job in managing energy use. He must; it is part of his feeling of well-being. In my experience, he rarely does as well as he might. The responsible person(s) simply does not have the data to control energy use in the face of the many variables encountered in day-to-day operation.

The traditional approach has been to compare current energy use (the audit data) against history. This is usually calculated as energy per unit production and tracked with time. This approach was fostered by the need to report figures as energy conservation progress to the government. More or less arbitrary savings targets (typically 20% for 1980) were set against the base year 1972. As shown earlier (see Fig. 1-1), these targets were grossly inadequate from an economic standpoint. This approach was also inadequate for control for another reason: it was not based on technical limitations, merely on whatever had been accomplished in a period of cheap energy.

Each plant needs an "evergreen" energy management system. Ultimately the system should have the characteristics shown in the following list.

Essentials for Energy Management Systems

Represents the best attainable operation at each plant
Has a solid technical basis, not historical
Includes adjustments for important variables
Helps identify corrective actions
Is suitable for frequent monitoring by operations
Acts as a basis for financial forecasts
Plays a part in the plant optimization program

Its objectives are to provide tight, technically based targets for energy consumption for the full range of expected operating variations to assist in day-to-day control and to provide a basis for financial forecasts and stewardship. There should be a target for each major piece of equipment to assist trouble-shooting. The economic impact of deviations from target should be calculated to prioritize remedial efforts.

Stewardship and forecasting are a natural offshoot of such a system. Predicted consumption can be compared against actual values in detail. With enough detail in target calculation and audit data, the reasons for variations should be easily understood.

A. Technical Targets

A system of technical targets that provides enough data for day-to-day monitoring of energy consumption and indicates the incentives for remedial action is required. Such a system can be expanded quite easily to provide a standard against which existing facilities can be measured and hardware improvements evaluated.

Figure 5-10 shows how such an energy target system would work. The chief element of this system is a target which is established by detailed technical analysis of the existing facilities. This target energy index represents the expected energy consumption of an existing plant when operated year-round at the best possible energy efficiency, i.e., the best achievable operation without additional investment. Since throughput, ambient conditions, feed quality, etc., will vary throughout the year, all of the major variables which affect energy consumption on the individual pieces of equipment must be accounted for in parametric studies. The difference between the actual operating data points on any piece of equipment and the target for the conditions under which the equipment must presently operate gives the incentive for improved operation on that particular equipment.

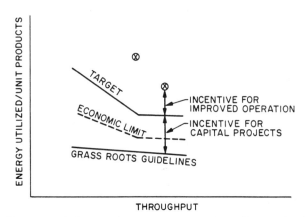

Fig. 5-10   How energy indexes work: an energy target system. ⊗, operating data points.

The sum of all incentives for all pieces of equipment represents the incentive at any moment for the entire plant.

The frequency with which this analysis can be made and remedial action taken obviously depends very much on the amount of technical and operating manpower available and the computer tools they have at their disposal.

To demonstrate the level of sophistication that is ultimately possible, we shall consider an example from a major plant. Figure 5-11a shows a typical debutanizer tower system in a modern ethylene plant. The tower is first simulated at design conditions with design quality tools. It is generally prudent to verify the calculated performance with a plant test under clean conditions. Parametric studies are then run with the verified model to quantify the impact of variations from design. After this an operating model is developed. It should be suitable either for building into the control computer, off-line processing by computer, or hand calculation. Figures 5-11b–f show the results of such a study for the reboiler of a debutanizer in graphic form. The first curve shows the reboiler duty, but the others, which are functions of changes in one primary variable, are plotted against a correction factor for the base-case duty. As shown, both the reboiler and condenser duties can be plotted on the same graph for a number of variables. All told, there were a series of 12 graphs and associated computer models that would allow calculation of the reboiler duty, the condenser duty, and the required reflux rate for the tower under a variety of conditions. The variables taken into account were throughput, overhead purity, reflux drum pressure, bottoms purity, available feed preheater duty, and feed quality.

Table 5-3 shows a comparison of the base-case reboiler duty and an arbitrary example chosen for off-design conditions. As shown, deviations in

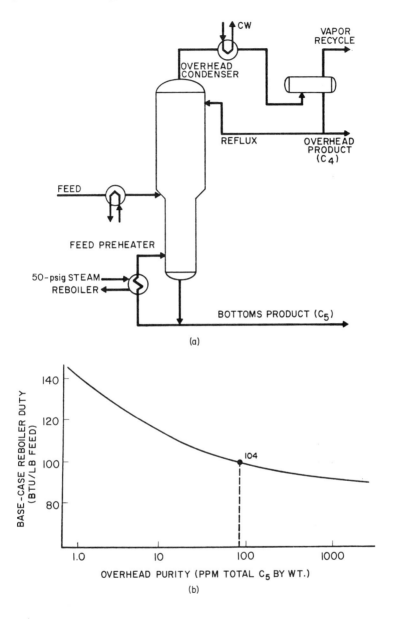

(a)

(b)

Fig. 5-11 A debutanizer tower system in a modern ethylene plant: (a), process diagram; (b), base-case reboiler duty versus overhead purity; (c), reboiler duty pressure correction factor versus reflux drum pressure; (d), reboiler and condenser duty correction factors versus bottoms purity; (e), reboiler and condenser duty correction factors versus preheater duty; (f), first-feed quality correction factor to reboiler and condenser duty.

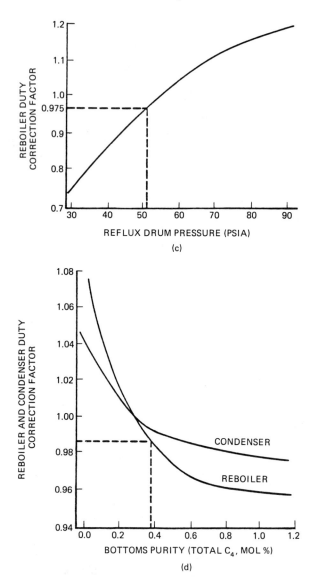

Fig. 5-11   (c) and (d)

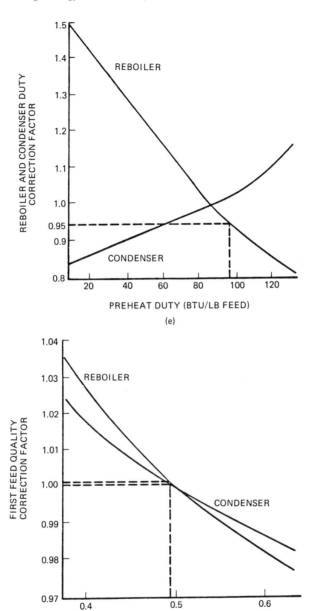

Fig. 5-11   (e) and (f)

Table 5-3

Calculation of the Corrected Reboiler Duty for a Debutanizer

Parameter	Base case	Example	Change (%)	Correction factor
Feed quality	0.508	0.5	−1.6	1.002
Feed rate (kpph)	112.3	112.3	—	—
Overhead pressure (psig)	55.1	52.5	−5.3	0.975
Overhead purity (C_{5+} ppm)	200	100	50	1.043
Bottoms purity (mol % C_4)	0.33	0.4	20	0.988
Feed Preheat duty (Btu/lb feed)	89.7	100	11.5	0.95
Reboiler duty (Btu/lb feed)	97.9	93.6	—	—

feed quality, overhead and bottoms purity, overhead pressure, and available preheater duty were taken into account with a net target of 4.4% less energy than the base case. A little further analysis shows that energy consumption for this tower is much more sensitive to preheater duty and overhead pressure than it is to feed quality and bottoms purity. A plant engineer will quickly reduce the time it takes to analyze the debutanizer operation to just a few minutes as he gains experience. In this way a target can be calculated for each plant system daily or each time a significant variable changes. With these sophisticated tools, remedial action can also be indicated and financial impacts developed to help set priorities.

Much of the benefit of the system can be obtained without a process control computer. The frequency of analysis is obviously reduced, and more technical manpower is required. Using these five graphs and calculating by hand, it took less than 15 minutes to determine what the reboiler energy consumption should be.

B. Targets Are Prepared by Cooperative Effort

In developing the plant target index, full participation by the plant organization is critical. Without this the required credibility within the operating ranks will not exist, and the technical personnel will be regarded as either foolish academics or emissaries from "Big Brother." In either case they will not be welcomed, and their objective of increased profits for the company will not be achieved.

The targets require detailed analysis of each major piece of equipment and its auxiliaries. As discussed earlier, a distillation tower with its condenser, reboiler, pumps and piping, and any feed or product collection – treating facilities would typically be grouped. Once a base-case set of operating conditions is established (perhaps through plant tests), operation at other conditions is simulated. For a distillation tower, for example, parametric studies involving feed and reflux composition, feed tray selection, throughput, cooling water temperature, ambient temperature, tower pressure, feed temperature and percent vapor, product purity specifications, and perhaps the entire steam balance would be undertaken. Obviously, computer simulations are a great advantage in making such calculations, but small programmable calculators are often quite suitable for some equipment. To develop the target index for a unit or an entire plant, the consumptions for each individual equipment block are added up at corresponding conditions.

The engineering manpower required to develop such a target system is appreciable. For a new ethylene plant, about $3\frac{1}{2}$ man-years were spent developing the technical input to the target system. Added to this was the manpower necessary for programming the technical information into the monitoring system particular to the plant.

C. Rapid Payoffs Are the Rule

Although the development of the operating target system is a big ticket item, major short term savings have often been immediately realized.[2] In one plant with an absorber depropanizer–debutanizer setup, much more absorbent was being circulated than calculated to be required for the purity specifications listed. Stream analysis verified that the purity from the towers was indeed exceeding specification by a wide margin. Conversation with the plant people showed that a downstream polymerization unit preferred the purer material to avoid minor operating problems. When the practice was instituted, energy cost 20¢/MBtu and the economic tradeoffs were correct. At $2/MBtu the energy required to produce this minor benefit downstream was no longer being wisely expended. As a result, a return to the original specification was implemented with a net savings of over three quarters of a million dollars per year (in 1979 dollars).

Opportunities to improve current efficiencies are often discovered by using the models developed for the target calculation to explore other methods of optimizing operating conditions. Adjustment of the pressures maintained in intermediate-pressure steam mains might be indicated. For example, if the pressure of the 616-psia steam main in Fig. 4-4 were reduced a little, less steam would have to flow through the condensing portion of CT-2. If no other changes were made, more steam would have to flow through the small drivers PT-1 through PT-5 that impact lower pressure levels. Because CT-2 is much larger than the others, this might be a good trade-off. Parametric studies with the steam balance model developed for the target calculation should quickly answer this question.

Compressor and tower studies often lead to similar guidelines, particularly at low throughput. If a multistage compressor is running in recycle to avoid surge, the energy model can show whether the recycle system is optimum for all stages. In one petrochemical plant the desired margin above surge was 10%, but to achieve 10% on the stage closest to surge a 30% margin was needed on the stage furthest from the surge point. Once this became apparent, independent surge protection was arranged for each stage, and 4% of the total horsepower in the machine was saved.

Reducing tower pressure levels can have equally beneficial results if the separation systems can accommodate the change. Energy savings accrue because less reflux is needed for most separations at lower pressure (better relative volatilities) and less compressor and pump horsepower is required. By using the energy models for all tower systems in concert, the optimum operating pressures for various throughputs and feed purities can be determined.

Ideas like these are often suggested by plant technical people but rejected by operations supervisors. Either the grass roots calculation effort is too large a distraction for the plant engineers, or the risk associated with a plant test is judged too great. However, when verified energy models have been developed, the plant management is generally more receptive. As the technical staff gains an understanding of their units through the use of the models, more and more practical ideas based on plant data are suggested.

Better day-to-day monitoring of operations can also save from 2 to 10% in a well-tuned operation that is subject to variations in processing requirements or ambient conditions. In winter refrigeration systems can take advantage of colder cooling water to reduce the discharge pressure of the compressor and save horsepower. In a large ethylene plant in Japan, optimizing the discharge pressure of the propylene compressor as a function of changes in cooling water temperature saved over $500,000/yr.

Often many small savings are possible if the operators can be provided with enough technical information to implement them. Ideally, a report can be provided each morning (or shift) showing each piece of equipment that is not operating at the target value calculated for the current operating conditions. An example of a very complete hypothetical report is shown in Table 5-4. Generated on the "management by exception" principle, it lists only pieces of equipment which are off target, showing the cost impact of the deviation and the plant data related to the problem. It is the type of report easily prepared using a process control computer but also possible by off-line computation.

The report provides enough information to begin to track down corrective

Table 5-4

A Hypothetical Operations Supervisor's Morning Report Showing Exceptions to Energy Targets

Variable	Target value	Actual value	Annual cost ($1000/yr)	Supporting plant data
Steam superheat temperature, 1 furnace (°F)	910	880	300	Desuperheater flow = 2.2 kpph, stack temperature = 340°F, O_2 at bridgewall = 1.8%
Furnace stack O_2 at bridgewall, 1 furnace (%)	1.8	2.8	25	Draft at bridgewall = 0.2 in H_2O
C_2 splitter ethylene purity (%)	99.975	99.990	100	Reflux rate = 890 kpph Overhead purge = 3.0 kpph C_2^- in bottoms = 0.6 mol %
Propylene refrigeration compressor surge margin (% above surge)	20.0	30.0	225	Suction pressure = 15 psig Kickback = 20 kpph
Blow-down rate, 1 furnace (%)	5.0	8.0	40	Check boiler feed water analysis

action in order of profit priority. Consider the steam superheat temperature. Clearly, the desuperheater water flow (DSH flow) is too high. The operator can attack this problem immediately. If maintenance is required, the cost reduction potential for the repair clearly sets its priority. Similarly, the key variables to explore are listed for all of the other items except for the blow-down rate. Since boiler feed water quality is often measured periodically, it may be necessary to develop more data before action can be taken. If corrective action cannot be taken immediately, the plant manager should at least know how much the deviation is costing him.

D. Use in Run Planning

Precise information on energy costs as a function of throughput can be used to eliminate fuzzy thinking in run planning. Figure 5-12 shows total plant energy cost as a function of throughput for two different types of plants in the United States. In a period of slow market demand, management had effectively mitigated the impact of less-than-capacity operations on the polymer plant by running the plant at full capacity until inventories built up and subsequently shutting down for periods of time. This operation was a semi-batch process, so startup procedures were well optimized. The logic for this plan of operation is shown by the steep slope of the energy versus throughput curve.

A similar approach was also proposed for the solvents plant, whose energy

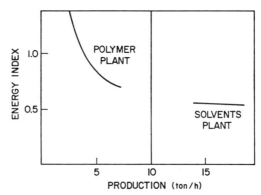

Fig. 5-12   The use of energy indexes (fuel oil equivalent tons/ton of product) in run planning: total plant energy cost versus throughput.

versus throughput curve is shown at the right in Fig. 5-12. This was a continuous operation with many stages, and startups were always a major event. The instinctive antipathy of the operating people to plant shutdowns was being ground down by the hard profitability data from the polymer plant until the energy curve was developed. The difference between the two operations clearly showed that generalizing from one plant to the other was folly. When other credits, such as somewhat higher recovery at lower throughput and startup losses, were worked into the equation, the plan for controlled shutdowns of the solvents plant was dropped.

E. Overall Optimization

One must never lose sight of the fact that energy costs represent only a part of total plant operating costs. Trade-offs between throughput and energy must often be made, and many times the best energy efficiency corresponds to conditions which are either difficult to achieve or involve higher maintenance costs. The obvious answer is to develop a sophisticated overall plant model that takes into account not only energy but also product values, feedstock efficiency, service factor and maintenance impacts, etc. Thus, the plant energy model can be part of an overall plant optimization model coupled with other models that represent important cost factors. Steps in the direction of overall plant optimizers are already commonplace in sophisticated petrochemical plants.

The target index studies have had some interesting fallout. The greater quantitative understanding of the operation, which is inherent in the study, provides tools that can be used for debottleneck studies, plant trouble-shooting, EPA permits, loss studies, and a host of other applications. When the plant technical organization begins to work with the tools provided for the energy study, they usually find many ways to use them in dealing with unrelated process problems.

F. Grass Roots Guidelines

We return now to the "grass-roots guidelines" shown in Fig. 5-10. This guideline represents the energy consumption per unit of product in a grass roots plant designed according to the best process technology and built to current low hurdle rates for energy conservation projects. This generally means that incremental investment is justified if the return on the investment matches the cost of capital to the investor. Many corporations use hurdle rates on the order of 15% discounted cash flow based on escalating fuel values for energy conservation projects.

The grass roots guidelines are prepared in sufficient detail so that the process sequence and individual hardware constraints of existing facilities can be compared against what would be justified in new grass roots situations. By making this comparison, ideas for facility improvements and the incentives to justify the required investment can be quickly developed.

Since retrofit investments to improve energy efficiency are often more costly than grass roots incremental investments, it is generally true that economic considerations will prevent the complete updating of existing facilities to grass roots efficiency. Thus, an "economic limit" is shown between the best operating efficiency of existing facilities and the efficiency of a new plant. This economic limit will change as the cost of energy changes, but it does represent a planning goal for corporate management.

Grass roots guidelines are largely a product of technical and/or R&D organizations. They represent the "state of the art" for proven technology worked out in enough detail to allow specific comparisons between the "best" and existing equipment. They need to be updated every few years as processes and equipment technology evolve. In general, the manpower required to develop these targets is relatively small, from 2 to 4 man-months, for a typical petrochemical unit.

For cases in which a major process change has occurred (such as the Union Carbide Unipol process), it may be necessary to develop two grass roots guidelines to complete a competitive assessment. The first would be for the process currently employed in the facilities and the second for the new process. Often this analysis triggers ideas for process development as well as serving as a standard against which the modernization of existing facilities can be analyzed.

Brainstorming sessions to develop ways in which existing facilities can be made to approach grass roots efficiency are often profitable. From these have come ideas for cascade reboilers retrofitted to existing distillation towers, heat pumps to eliminate the condenser heat waste altogether, and intermediate reboilers to save an expensive utility at the cost of waste heat or a similarly inexpensive utility. Projects including all of these ideas have been profitably implemented in modern petrochemical plants.

G. Implementation Faces Some Roadblocks

The following list describes some impediments that can only be overcome by diligent technical and management efforts.

Fears and Roadblocks to Implementation

Won't really reflect plant situation
No margin for essential variations
Comparison with other plants
Too much work
No computer

The first two items represent technical problems and are the easier of the group to deal with. If targets are largely being developed by members of the technical staff, the assignment of one or more plant engineers to the team will often get around the natural resistance to accept standards imposed by outsiders. In addition, this should build up a ground swell of credibility in the plant organization. This is more easily done if carefully chosen plant tests are used to validate the model and some *immediate* short-term efficiency improvements are identified. It is particularly important to work out the impact of variables the plant people feel are important. This is often a little cumbersome to deal with, but generally worth the effort.

The larger problems are associated with developing the proper management environment. Local plant management often has a justifiable pride in the efficiency of its operations. The first reply is, "There is nothing to be found here, we run a tight ship." Invariably this is not correct, however sincere the statement. Even plants with existing computer control facilities have been found to profit from the utilization of more detailed technical models for their equipment.

The real problem comes from the paranoia of local plant management that central management will use the data to make life miserable for them. The temptation for the manufacturing vice president or the product line vice president to do this is very large. When advised that operating costs can be cut without additional investment, the first reaction of many a vice president is something like "those sloppy you-know-whats." This, of course, is not true. The plant management never had the benefit of such sophisticated analysis before. *Recrimination, even in recession times, is the worst possible approach.* Top management will find much more money in its cash register if it adopts a positive attitude along the lines of, "Now that you have all this good data show me how you are going to improve profits with it."

The direct comparison of similar plants is also a temptation that top management must avoid. Actually, the details available to the plant manager should give him plenty of ammunition to shoot down any unfair comparison. If fair comparisons should be made, the plant manager is in the position to make them and take action on required changes long before they are brought to any vice president's attention. All of this is not immediately perceived by management, however, and the leader of the target program must tred carefully and seek positive results early if this natural paranoia is to be circumvented.

H. Computers Help but Are Not Essential

If the plant has a powerful computer capability, it is in a better position to implement and utilize the target program. The parts of the target system directly related to controlling the operations can be put into the process control computer, while other parts are done by off-line analysis. However, the key element is not the sophistication of the tools in place but the extent to which they are *used* by decision-makers. Daily off-line use of simplified tools by the operating supervisors will save the company a great deal more than sophisticated hourly calculations that are paid lip service by supervisors or used to explain in great detail each month why targets are not met. The operating target system must be tuned to what the plant can, and will, use frequently. The temptation for useless (and costly) technical elegance must be resisted.

Summary

Adequate technical information can be used to prevent waste and to provide an almost unlimited degree of sophistication in daily energy management. Significant improvements in plant energy efficiency have been achieved with little or no investment by giving plant organizations the important technical data they need in a usable form. Control of excess oxygen in combustion and elimination of leaks in steam traps and systems are two pervasive areas of opportunity. Technical target systems for energy use, which define the best a plant can do at a variety of throughputs and process conditions, have also been used to identify opportunities and guide prompt responses to changes in operating conditions. Both activities can contribute significantly to current profits (5 – 10% savings). In addition, the data provide a sound base case against which to evaluate investment ideas. In many cases the technical understanding gained can also be used for other purposes such as debottleneck studies and run planning. The key element is to *use* whatever information can be had, rather than waste time on useless technical elegance. The following list summarizes the several ways that energy target studies can be used to improve profitability.

Sources of Benefits from Target Program

Grass Roots Guidelines

Identify investment opportunities
Assess competitive situation
Evaluate technology for new investments
Develop basis for development efforts

Targets

Removing historical nonequipment constraints
Monitoring of day-to-day operations
Improving run plans
Providing input to overall plant optimization program
Providing fallout for other plant studies

Notes to Chapter 5

1. S. J. Vallery, Steam traps: the quiet thief in our plants. *Chemical Engineering* **88,** 85 (1981).

2. W. F. Kenney, Monitoring Energy Efficiency in Sophisticated Process Plants. Proceedings of the 3rd Industrial Energy Conservation Technology Conference, Houston, Texas, April, 1981.

6

Facilities Improvement
An Overall Site Approach

Introduction

Before launching into the subject of this chapter, let us pause a minute to recap. The first two chapters struggled to get across the idea that the world supply of oil is decreasing, which will likely lead to an upward trend in its price, and that the popular word "energy" is very inexact. We learned that the various forms of energy have different values and that our task in energy conservation is to maximize the desirable change that is achieved from every Btu of "available energy" used. We then proceeded to characterize the types of energy that are used in a plant and to launch into the two noninvestment phases of an energy conservation program, namely, minimizing waste and managing energy use. Having developed a set of energy values for the utilities supplied to a process, we are now ready to discuss improving the facilities at hand. All of this knowledge is essential to establishing a proper base case against which to evaluate the economics of investment opportunities. A nonoptimized, misunderstood base case can lead to wasteful investment decisions and even to the unemployment compensation line.

Just as noninvestment options fall into two categories, so do investment opportunities. The first step is to maximize *intraunit conservation* within the guideline of matching suitable sources of energy to requirements. This is generally the most economic approach, because extensive transport systems are not involved and no redundant or backup investments are required to insulate the service factor of one unit from the next.

The second phase of the facilities improvement effort involves reworking the energy system to minimize the consumption of available energy while supplying the necessary process requirements. In effect, this means that the *interaction* of energy conservation measures taken at various sections of the plant is considered in such a way so as to optimize the overall available energy consumption in the plant. After all, it is the profit of the entire plant, not a single unit, that we are striving to improve.

I. Utilizing the Energy Audit

A. Brainstorming Procedures

The data generated in Chapter 4 will be used to demonstrate a systematic procedure for identifying ways to improve onsite facilities. Figure 4-2 (p. 63) defines the problem. The reader is asked to play the role of the plant energy conservation engineer charged with improving facilities to reduce energy costs. In fleshing out the energy audit in Chapter 4, Tables 4-7, 4-8, and 4-9 were developed to characterize all of the energy inputs and outputs of the process. These tabulations will assist in identifying opportunities to improve the facilities.

The first step in using the audit data to improve energy efficiency is to evaluate whether all of the noninvestment steps have been implemented. In this case the answer is a loud NO! Ten percent oxygen in the stack corresponds to over 50% excess air, which can be significantly reduced. Indeed, the savings associated with facilities improvements cannot be estimated until this correction is made. Very roughly (from the available heat diagram in Fig. 5-5), the efficiency of the furnace can be increased to 77% by reducing the level of excess air to 20%. This produces savings on the order of 3 MBtu/h without any significant change in facilities, and 20% excess air becomes the base case.

Having established a correct base case, we are now ready to suggest ideas for updating the design. In general, these ideas will be developed in a brainstorming session involving people knowledgeable in the process, those knowledgeable in energy conservation techniques, and perhaps some additional people noted for their creative thinking. On major units it is sometimes profitable to bring in an outside "facilitator," either from within the company or from a special consulting firm that provides this service. Whether this is done formally or not, it is very important to make sure that no judgment or evaluation of ideas takes place during the brainstorming session. Each idea and its potential is simply written down without critical review. The key is to build participation during the session, to enhance the feeling of freedom. This approach generally leads to more and better ideas and to combinations of ideas with synergistic effect.

Another valuable approach for ensuring that no ideas are missed is sequencing. This involves the discipline of polling each participant in order and limiting each to one idea per turn. This device prevents experts from inhibiting others, averts domination by a few strong personalities, and ensures that a timid soul will get a chance to speak his piece. As is often the case in such brainstorming sessions, ideas not related to the original main objectives may surface. Save these also. Take some time to write down your suggestions for improving the particular process shown in Fig. 4-2.

The procedure for idea generation involves two steps:

(1) a systematic attempt to match energy needs with rejection steps (sources and sinks);

(2) a reevaluation of every assumption inherent in the orignal process and equipment designs.

In most cases several alternate ideas are possible, and the energy conservation program must select the most economic of all of these possibilities. However, the first step is to identify *all* of the possibilities for improving the facilities.

B. Matching Sources and Sinks

Some focus for brainstorming on heat integration possibilities can be developed from a study of Tables 4-8 and 4-9, which list energy inputs and rejections along with process requirements. This is often helpful in the early stage of idea generation. For the problem in question, three major points of attack can be suggested:

(1) improve furnace efficiency;
(2) recover heat rejected to cooling water in major heat exchangers;
(3) save steam in the preheater and at the pump turbine.

The next step should be to *list all of the ways possible for attacking these three objectives, along with the potential for improvement that goes with each idea.*

Five ways to improve furnace efficiency suggest themselves. From the available heat diagram, we see that furnace efficiency could be improved to about 93% if stack temperatures could be lowered to 300°F. Table 4-8 shows that feed preheat could be carried out with an additional convection section. This might provide most of the required heat sink and matches the temperature well. Of course, both of the "standard" approaches to furnace efficiency improvements, combustion-air preheating and steam generation, can be considered. In addition, temperatures are such that the current reboiler duty could be served with less fuel firing if a larger convection section were installed. However, the minimum stack temperature in this case would be about 400°F. These items are listed in Table 6-1 as Nos. 1A–1D.

Table 6-1

Summary of Potential Waste Heat Recovery Ideas for the Unit

Idea	Potential savings (MBtu/h)
Operational	
1. Cut excess air to furnace	3
Investment	
1. Improve furnace efficiency	
A. Preheat feed in convection section	13
B. Add waste heat boiler	15+
C. Preheat air to 700°F	8
D. Add more surface for present duty	10
E. Generate power via Rankine cycle	3–4 (as power)
F. Substitute steam heater and eliminate pump	$\Delta_{eff} \times 38$
2. Recover heat rejected to cooling water	
A. Reuse toluene cooler heat to preheat feed	5
B. Reuse part of condenser heat to preheat feed	5–7
C. Preheat furnace air to 150°F – 200°F	1–2
3. Save steam at turbine	
A. Replace with motor	
B. Improve efficiency (machinery or condenser)	
4. Add heat pump on tower	
5. Substitute another separation process: crystallization	

In a more costly vein, the waste heat could be recovered to generate electricity via an organic Rankine cycle (1E). Here, the principle of trying to upgrade the quality of the heat recovered by recovering a form of energy that has inherently higher available energy content is being practiced. The amount of power generation possible holds the key to the economics of this idea.

For recovery of rejected heat, the field is more limited. The only real use for lower-temperature heat is in feed preheating or low-temperature air preheating. The toluene cooler E-2 and/or part of the condenser E-3 might be used to save at least 5 MBtu/h of steam by preheating feed (as approximated from $C_p M \Delta T$ using data from Chapter 4). This heat might also be used to save fuel directly by preheating combustion air to 150–200°F (depending on the exchanger used). These items are listed as 2A–2C in Table 6-1.

Before completing the list, the turbine-driven recirculation pump must be considered. The theoretical fluid horsepower was calculated from the flow and head to be 72 hp in Chapter 4. The theoretical steam rate should be 954 pph instead of the measured 3400 pph. This corresponds to an overall efficiency of 28%. A motor or a more efficient turbine should be considered (Nos. 3A–B).

The turbine could also be discharged into a condenser preheating the feed at the cold end. This would lower the turbine back pressure, recovering more work per pound of steam. Also, turbine efficiency would be less of a factor, since all the exhaust heat would be used, albeit at a low level. Perhaps you can add other ideas to the list.

C. Question Assumptions for More Potential

To generate additional possibilities, it is necessary to step back and question why the current design is as it is. *Each assumption inherent in the process sequence needs to be reexamined.*

Let us plunge into the deepest water by asking, "Why fractionate at all?" The question is more than trivial even for this simple process. It sets the standard for the extent to which this reexamination should go. *Nothing is sacred!* It also sets the limits of tolerance for new ideas: *Infinity!* At this stage there should be no rejection or evaluation permitted.

Specific suggestions are still required, however. What could be used instead? Crystallization is one possibility. It is listed as No. 5 in Table 6-1. Other thoughts may occur and should also be recorded pro tem, in spite of the practical opinions of anyone present.

If we accept distillation as the process to recover these products, there are still assumptions to be questioned. Specifically, is it necessary to put heat into the bottom of the tower and throw it away at the top? Is it necessary to use so much reflux? Is it necessary to use a furnace as a reboiler? Are the process conditions suited to current energy costs? Are the product quality specifications still appropriate? These questions can be used to generate other ideas.

To demonstrate this let us consider the reboiler question. Clearly, a steam heater is technically feasible. Steam at 150 psig condenses at about 365°F. This would give a mean temperature driving force of 50°F. In addition, a thermosiphon design might be used, obviating the need for a pump. This idea is listed as No. 1F.

More fundamentally, an engineer experienced in distillation schemes or energy conservation technology might suggest converting the tower to a vapor recompression cycle. In this cycle the overhead vapors are compressed to such a pressure that they can be condensed at a higher temperature than that required to supply heat to the bottom of the tower (see Fig. 6-1). Thus, the latent heat of the distillation operation is recycled by substituting a smaller amount of high-grade energy in a compressor for the much larger amount of energy currently used in the reboiler. (At present that reboiler also uses high-grade energy.) The existing reboiler and condenser can be used for startup and trim cooling, respectively. This idea is listed as No. 4.

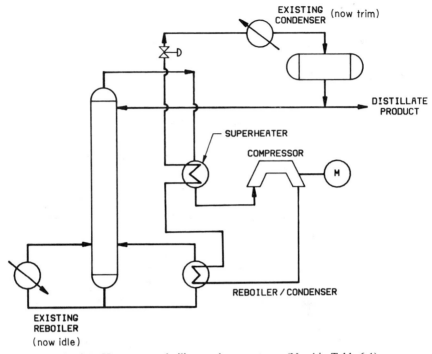

Fig. 6-1 Heat pump reboiling on the same tower (No. 4 in Table 6-1).

Other ideas for changing process conditions might also be suggested. Without additional data such as the detailed tower balances, tray-to-tray calculations, and tray hydraulics, however, no specific idea can be offered. The brainstorming team should pursue these items before completing the list.

Hopefully, this example has demonstrated both how the audit data can be used and the value of having diverging skills in the brainstorming group. The key elements of the process are restated in the accompanying list for emphasis.

Before moving on to some preliminary evaluation of the idea list, some comments on its character are in order. Note that some of these ideas are self-contained within the unit. These include combustion air preheat on the furnace reboiler and the heat pump, if a motor driver is used. The other ideas involve interaction with the steam system. Reducing consumption of 150-psig steam might be economic or might not depending on the steam balance in the plant. Increasing the total 150-psig steam demand also cannot be evaluated unless the entire steam system is considered. If a combination of changes to the onsite steam utilization could be developed which leaves

---

A Systematic Approach to Identifying Improvements

Prepare an energy audit and requirement – rejection tables
Assemble a brainstorming team of diverse background
Search for improvements that do not require investment
Initiate a two-stage search for design improvements
    Match sources and sinks within the unit
    Reexamine assumptions in process and design
Do not evaluate or reject at this stage
Simplify the list of ideas qualitatively
Explore site energy system interactions, consolidate list
Evaluate overall site economics

---

the total steam demand about the same but reduces or eliminates the fuel fired in the furnace, this too could be a self-contained energy efficiency improvement. Indeed, if the site steam system is in reasonable balance to begin with, a self-contained idea can progress more quickly because it enhances the thermodynamic efficiency of the process without causing site interactions. Since steam balances often change from summer to winter, partial-year credits may need to be considered in evaluating steam demand changes in detail.

This brings us to the question of the interaction of site energy systems. Before exploring this more complex issue, the idea list can be simplified by qualitative considerations to minimize downstream work.

D. Simplify the Idea List

In many cases equal savings can be obtained by different methods. Obviously, the least costly would be the best. Qualitative comparisons of capital requirements can often identify the most attractive method without extensive cost estimation efforts. From a practical standpoint this is desirable, because often many ideas are generated and they become expensive to evaluate in detail.

Consider the list in Table 6-1 from this point of view. Two sizes of heat recovery projects are described: a small one for about 5 MBtu/h and furnace heat recovery for 10 – 15 MBtu/h. Of the smaller possibilities, use of the toluene cooler probably requires the least capital. At the cost of some piping, feed can probably be substituted for the cooling water and most of the heat can be recovered directly. Of course, the toluene will exit at a somewhat higher temperature than the present 120°F, unless the existing exchanger is oversized. A small trim cooler may be needed. Use of the condenser heat will

Table 6-2

Equipment Required by Projects 1A, 1B, 1C, and 1E of Table 3-5

	Project			
Equipment	Preheat feed (1A)	Make 150-psig steam (1B)	Preheat air to 700°F (1C)	Generate power (1E)
Heat exchanger	Base	≥ Base	> Base	≫ Base
Flue-gas ducts	—	Large	Large	—
Forced-air fan	No	No	Yes	No
Induced-draft fan	Yes	Yes	Yes	Yes
Boiler feedwater supply	No	Yes	No	No
Vapor drum	No	Yes	No	No
Foundations or structure	Base	> Base	> Base	≫ Base
Electrical supply	Base	Base	> Base	≫ Base
Machinery	No	No	Minor	Major

probably require a greater investment, because the existing exchanger must be retained in cooling water service to handle the bulk of the duty. As a result the entire preheat duty must be provided in a new exchanger. Thus, project 2A (toluene cooler) should be compared first with the larger projects.

For furnace efficiency improvement, note that projects 1A, 1B, and 1C recover about the same amount of heat and that 1E uses this heat to recover a lesser amount of power. Very roughly, the facilities needed to carry out each of these projects can be compared to shed some light on relative capital costs (see Table 6-2).

On this qualitative basis, direct feed preheating appears to be the best possibility, because capital costs would be low. Slightly more heat recovery may be possible with the waste heat boiler, because not all the available heat in the flue gas is needed to preheat feed; however, incremental capital costs are required. Alternatively, the air-preheating project has the benefit of directly reducing fuel, whereas the value of steam generated must be determined from the plant steam balance.

For electricity generation capital costs are generally in the range of $2000/kW. Qualitatively, this system has a higher cost for heat exchangers, machinery, and structure than any of the other cases and comparable costs in other areas. The attractiveness of this investment depends on the relative prices of electricity and fuel.

The plant program thus boils down to deciding whether both feed preheating and furnace efficiency improvements can be justified. The most likely cases are either 1A, or 2A and 1C, unless further process, layout, or safety considerations decree otherwise. These projects can be considered first in an effort to establish the overall attractiveness of the idea list before large-scale commitments to engineering and manpower are made.

The furnace replacement and heat pump projects deserve special consideration. The capital cost is likely to be high, because of the need for a large heat exchange surface and machinery. However, net savings are also likely to be large. To evaluate these possibilities correctly, there is a need to consider a number of site interactions and related costs. They cannot be evaluated quickly.

II. Overall Site Interactions

Let us hypothesize a site energy system to go with the unit being discussed. Consider the site energy system in Fig. 6-2, which shows the furnace as a rather isolated fuel requirement 1 mile from the main boiler plant. The steam demands of the process unit are included in the process heating block shown near the furnace along with those of other units. The key element of this energy system is that there is a constant power demand independent of the unit to be improved. The pressure on the 150-psig steam system is controlled by condensing excess steam in a dump condenser. In this system the boiler must always be fired at the rate shown to produce the constant power demand.

The inherent inflexibility of the steam power system impacts on a number of the ideas that were suggested for improving the energy efficiency of the aromatics unit. If steam were to be generated in the aromatics unit or if the process-heating steam demand were reduced, this would simply force more steam into the dump condenser and save no fuel at all on the overall site. Thus, investments to make these changes will have no net return. Alternatively, increasing steam demand reduces the load on the dump condenser and credits come from reduced fuel fired at the aromatics site. Thus, replacing the aromatics furnace altogether seems to be more attractive.

Interactions with the site energy system also point the way to some new ideas. Clearly, the heat from the dump condenser could be used to provide some combustion air preheating at the boiler. A quick check shows that more than enough heat is available in the dump condenser to heat the boiler air to the pinch-point temperature ($\sim 300°F$). This possibility might be extended to the aromatics furnace and the site restored to steam balance. Heat from the aromatics furnace stack is available at higher temperatures than 150-psig steam. Thus, an additional idea would be to use a heat transport loop to move excess heat from the furnace to the boiler for

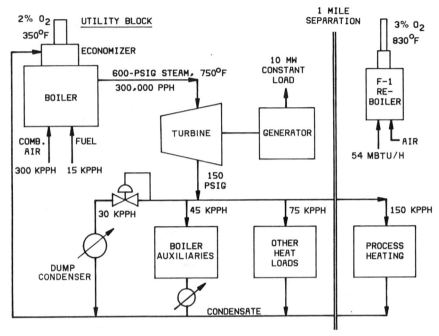

Fig. 6-2   Hypothetical site steam–power interactions.

high-temperature air preheating. This could be in addition to an air preheating project at the aromatics furnace itself.

Refineries have long engaged in the practice of using "hot oil belts" to move high-level (and thus high available energy) heat around the site. As the cost of energy rises, some of the old worries about reliability are being overcome by providing alternate sources and sinks. Manning[1] describes a system recently put in operation at a Shell refinery in Louisiana. As in any such system, the local operability requirements were important and some waste heat recovery was foregone to achieve a practical result. O'Brien[2] gives further insight into the potential economics of heat transport loops. From a thermodynamic standpoint, such systems provide the opportunity to recover more valuable energy, because they are not constrained to temperature levels dictated by the existing steam system.

To summarize the insights gained from considering site interactions, some additional constraints were imposed and two additional ideas were generated.

(1)   Generating or saving steam at the aromatics unit would not be

attractive because it would be wasted in the dump condenser. However, using some of the waste steam for boiler combustion air preheating might restore the site to a reasonable balance.

(2)   Some of the steam for replacing the furnace could come from that currently rejected in the dump condenser.

(3)   An idea for using high-level heat from the furnace to preheat combustion air at the boiler was generated.

Thus, our list of potential energy conservation steps has been narrowed down somewhat, and a renewed emphasis has been placed on supplying the correct level of available energy input to suit the fundamental process requirement.

The point of this discussion is not to discourage or delay onsite energy conservation improvements. *Rather, it is to point out that such steps can only be truly economic when the entire site consumption of available energy is reduced.* An insular approach can lead to some very embarrassing situations. In the simple system pictured here, an engineer can see his way through the interaction effects and choose among his onsite opportunities in a way that complements overall system requirements. As more complex systems are dealt with, intuition is no longer adequate, because the number of possibilities increases by orders of magnitude.

A. Qualitative Picture of Site Interactions

Qualitatively, a more general picture of some of the typical energy system interactions is given in Fig. 6-3. Essentially, work-related steps are pictured in the upper area; in the lower left area are the furnaces and boilers that power the steam systems, machinery, and process exchangers. In the lower right area are the desired processes. These require and/or liberate heat and power. Ultimately, all three sectors transfer degraded energy to the ambient sector, through which all interaction with the environment takes place. *The task of the energy conservation engineer is to derive as much useful work as possible in the entire system from the available energy inputs in the fuel and electricity before the rejected energy passes into the ambient sector.* To do this the engineer must capitalize on the interactions among the three other sectors so crudely pictured here as well as minimize the net use in each sector. The potential for complexity, and hence the challenge, is clear.

Various people have tried to devise systems to provide quantitative insights to energy and facilities planners about interactions at their site. Heuristic approaches are helpful, but specific applications still require meticulous analysis. Some guidelines will be offered at the end of this book, and their relation to available energy fundamentals will be unmistakable.

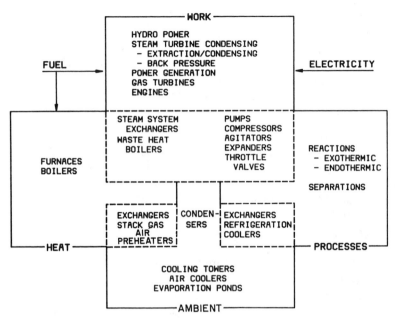

Fig. 6-3 Typical energy system interactions.

First, however, some background on why these rules are fundamentally sound will be developed.

B. Why Cogeneration?

Reiterating the task outlined earlier, we might say, "Recover as much available energy as possible at each level of need before passing what is left into the next lower zone of requirements. This precept sounds suspiciously like motherhood. Yet, there are a number of very specific things that can be done to achieve this. Clearly, the matching of appropriate energy sources to process requirements discussed in the aromatics example is one such step. Another is to *maximize cogeneration in the site interactions.*

The term "cogeneration" has become a popular "buzz word" for energy efficiency in recent years. This popularity emanated from simple Btu counting, but the correctness of the conclusion can be substantiated by thermodynamic analysis. In the discussions in Chapter 3 relating to the value of exhaust steam from a turbine, simple, balanced cogeneration systems were discussed. Figure 6-4b shows such a system. For the conditions imposed on the quality of the process heat, namely, an exhaust pressure of 200 psia from the turbine, *the cycle shown represents the maximum fuel efficiency possible*

Table 6-3

Ratios of Power to Heat at Maximum
Efficiency For Back-Pressure
Turbines

Turbine exhaust pressure (psia)	steam heat load / power produced
140	5.05
200	5.18
400	8.47
600	11.90

at the steam pressures selected. To balance the system and achieve maximum efficiency, 17,645 Btu/h of process heat must be used per kilowatt of electricity produced. As was shown in Table 3-1, if this process heat requirement is less, the same amount of power can only be produced if some steam is exhausted from the turbine (at very low pressure) to a condenser and the latent heat rejected to cooling water. This increases the fuel requirement (available energy input) per Btu of available energy actually used in the plant, because some at the fuel input must be rejected in the condenser.

Returning to the maximum efficiency case, we can calculate from the data in Fig. 6-4b that approximately 5.18 Btu of steam heat load is required per Btu of power for the example considered. The range of steam heat loads to power requirement ratios for the maximum efficiency cases at other turbine exhaust pressures is tabulated in Table 6-3. Clearly, it becomes more and more difficult to balance process heat and power loads as the exhaust pressure of the turbine is increased. On the other hand, there are more potential users for the exhaust steam because of its higher temperature (available energy content). If this higher available energy is not used to serve higher temperature (or pressure) needs, then an available energy loss will occur in the heat exchanger. Thus, we return to the principle of matching the quality of energy supplied to that needed by the process for maximum efficiency.

Consider the two approaches shown in Fig. 6-4 for providing 1 kW of electric power and 17,645 Btu of process heat as 125-psia steam. In the "conventional" cycle two separate boilers are used to provide high-pressure steam for power generation and 125-psia steam for process heating. Both are efficient in the first law sense, with 200°F stack temperatures. In the cogeneration cycle a single boiler is used, which raises much more high-pressure steam, but the exhaust steam from the turbine is used to produce the 125-psia steam for process heating. Because no heat is wasted in a turbine condenser, less fuel is required in the cogeneration cycle.

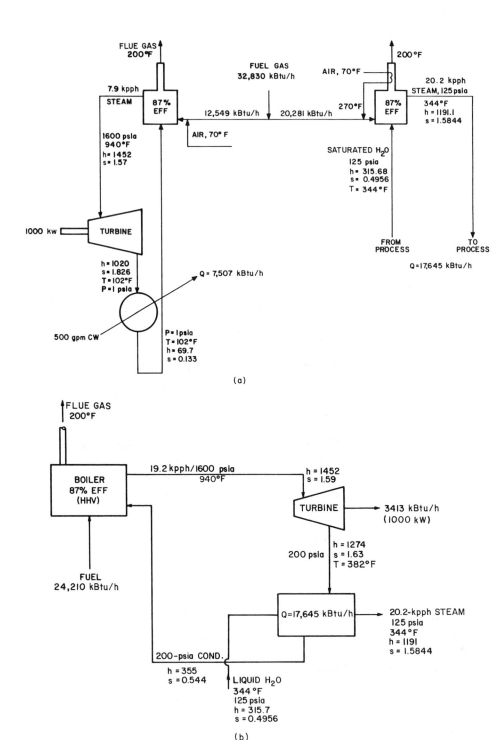

Fig. 6-4 (a) Conventional and (b) cogeneration cycles for a steam power system.

Table 6-4

Lost Availability Comparison of Conventional and Cogeneration Steam Power Systems

Parameter	Conventional system	Cogeneration system
Availability input (kBtu)		
Fuel at 21,370 Btu/lb[a]	29,362	21,648
Boiler feed water at 344°F and 125 psia	1,103	1,103
Total input	30,465	22,751
Availability Output (kBtu)		
Electricity	3,413	3,413
Steam	7,130	7,130
Total output	10,543	10,543
Total lost availability (kBtu)	19,922	12,208
Second law efficiency (%)	34.6	46.3

[a] Based on Sussman's "availability (exergy) analysis."

The direct relation of available energy and fuel has been discussed in earlier chapters. In the next chapter the methodology of thermodynamic analysis will be discussed in detail, but a simplified analysis is presented here (see Table 6-4) to reinforce the relationship between the amount of fuel fired and lost work (available energy).

The results are plain. The useful output of both systems is identical, as is the input from the 125-psia boiler feed water. The difference between the two cycles rests with the fuel requirement. Since combustion is a particularly irreversible process, more lost work is associated with the conventional process and the thermodynamic efficiency is lower. This little diversion provides some quantitative support for the popular affection for cogeneration in its simplest form.

The simple analysis of the two processes can be broken down to identify further the sources of lost work in each process. A somewhat more detailed analysis is given in Table 6-5. Combustion losses are proportional to the amount of fuel consumed. Steam generation losses are really a function of the temperature driving forces used in its production. The lower the steam pressure produced in a fired boiler, the greater the loss of available energy. This principle was discussed in general terms in Chapter 2, and here is one concrete example of the effect. Note that the work lost in heat exchange for the cogeneration cycle occurs in the 200-psia/125-psia boiler, and its contribution to the total lost work is only $\frac{1}{10}$ that of steam generation in the fired 125-psia boiler.

In this simplified preview of the insight provided by a thermodynamic

Table 6-5

Sources of Availability Losses in Conventional and Cogeneration Steam
Power Heating Systems

Source	Lost availability (kBtu)	
	Conventional system	Cogeneration system
High-pressure boiler		
Combustion	3686	7155
Steam generation	2604	3872
Low-pressure boiler		
Combustion	5958	—
Steam generation	6122	—
Turbine (net)	1072	615
Heat exchange	419	606
Total	19,861	12,248

viewpoint, the first fundamental guidelines for the engineer dedicated to
reducing his plant's energy bill emerge:

(1)   minimize combustion,
(2)   generate steam at maximum pressure and generate power from all
low-pressure steam demand (i.e., maximize cogeneration),
(3)   use steam for process heat at minimum temperature.

Note also the differences between the efficiency values and locations of
losses between a first law viewpoint and the one advanced here. First law
boiler efficiencies are listed at 87%, with the major waste at the condenser.
From the second law viewpoint, there is little lost in the condenser. Thus, a
very different energy conservation strategy is indicated by available energy
analysis compared to simple energy accounting.

The condensing turbine cycle shown in Fig. 6-4a is essentially the same
one used by utility stations. Utility stations are bigger, somewhat more
efficient, and probably use less expensive fuel than industrial plants. Therefore, the cost of power from their condensing cycle is probably less than that
from an in-plant cycle. The conclusion is, *eliminate in-plant condensing
turbines except those necessary for control.* This is the corollary to the earlier
rule to "maximize cogeneration." In practice electric motors are generally
better than in-plant condensing turbines both from an operating and capital
cost viewpoint.

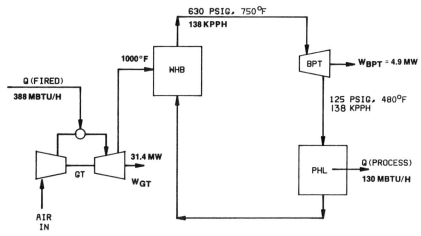

Fig. 6-5 Typical gas turbine–unfired waste heat boiler system. WHB, 630-psig waste heat boiler; GT, gas turbine; BPT, back-pressure steam turbine (75% efficiency); PHL, process heat load (125-psig steam). Process steam load/power generated $= 1/R = (130 \times 10^6)/(36.3)(3.413 \times 10^3) = 1.05$.

C. Gas Turbines and Diesel Engines

Steam turbines are not the only in-plant prime movers capable of being adapted to cogeneration schemes. Other driver types have different ratios of power to process steam requirement for maximum efficiency. Gas turbine drivers produce power directly from fuel and exhaust large amounts of combustion gases at temperatures between 1000 and 600°F. The lower temperature is generally achieved when the turbine is equipped with a regenerator to minimize its own fuel consumption. A typical configuration for a gas turbine driver would be to affix a waste heat boiler on the exhaust to generate 600-psig steam. In a combined cycle the 600-psig steam would flow through a power turbine exhausting at the pressure of required process steam. Such a cycle is shown in Fig. 6-5 for a 125-psig exhaust pressure. The ratio of process steam demand required to maximize the fuel efficiency of this system is 1.05 Btu steam/Btu of power produced. This is clearly lower than in the all-steam turbine case.

A diesel engine produces even more power per Btu of process steam demand required to maximize efficiency (see Fig. 6-6). This is because a diesel engine operates with much less excess air than a gas turbine and carries out combustion at a higher pressure. For a typical diesel engine the combustion gases exhaust at ~1000°F and the same type of waste

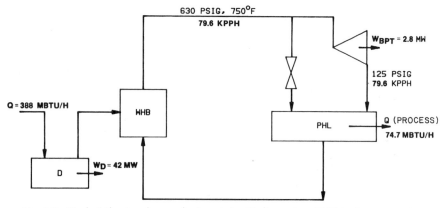

Fig. 6-6   Typical diesel engine–unfired waste heat boiler system. BPT, back-pressure steam turbine (75% efficiency); D, diesel engine; PHL, process heat load (125-psig steam); WHB, unfired 630-psig waste heat boiler. Process steam load/powered generated $= 1/R = (74.7 \times 10^6)/(44.8)(3.413 \times 10^6) = 0.49$.

heat boiler can be affixed. A typical ratio of required steam demand at 125-psig/kW of power for a diesel engine is 0.49. Both of these figures compare directly with a value of 5.0 for a steam-turbine-only system at the point of maximum fuel efficiency with 125-psig exhaust pressure.

III. Total Site Cogeneration Potential

It is possible to generalize this concept into relationships that are applicable to an entire plant's energy system. The purpose of this type of analysis is to provide more quantitative guidelines for the type of equipment most likely to produce maximum fuel efficiency in any plant and to define the extent of process heat demands required for maximum fuel utilization.

The principle that evolves from these observations is that the site energy engineer should maximize the power produced from the available process heat load. This becomes a problem in driver selection, but many other nonenergy constraints must also be dealt with.

The ratio of nonmotor power requirements to process heat that can be supplied by steam can be used to characterize the site cogeneration potential and to guide driver selection. Mathematically, this ratio is defined by

$$R = \frac{W(\text{Btu})}{[\Sigma \ Q_i]_{P_{\text{mean}}} (\text{Btu})}, \tag{6-1}$$

where W is the sum of all horsepower demands not supplied by electric

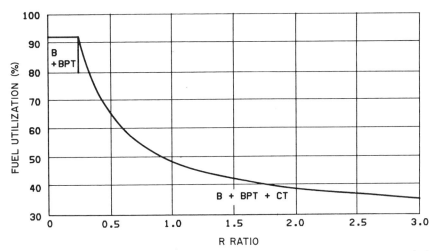

Fig. 6-7 Fuel utilization versus R ratio for a 125-psig process steam system at typical efficiencies. B, 1500-psig boiler; BPT, back-pressure steam turbine; CT, condensing steam turbine. R = total shaft work/net process steam heat load. Fuel utilization equals total shaft work plus net process steam heat load, divided by total fuel input.

motors and $[\Sigma\ Q_i]_{P_{mean}}$ the sum of direct-steam-supplied process heat demand at *weighted average process steam pressure.*

The maximum possible fuel utilization at the site is a function of the ratio R and the driver combination used. The fuel utilization is neither a first law nor a second law efficiency. It is merely a measure of the fraction of fuel energy that can be used *for something.* It is defined as the total shaft work produced plus the primary steam process heat load in Btu's, divided by the total fuel fired in Btu's. Process heat loads supplied directly by fuel and power needs supplied directly by motor are not included. Neither are electric heating loads. Also excluded are process heat loads supplied by multiple effect use of energy. For example, if the overhead vapors from a steam-heated tower or evaporator are used to heat another exchanger, the duty of the second exchanger would not be counted.

In Fig. 6-7, which is for steam turbine power systems *at typical efficiencies only,* we see that up to a ratio of about $R = 0.2$, power can be produced from a back-pressure turbine, the exhaust steam going to serve process heating loads. Beyond the ratio $R = 0.2$, any additional power must be produced by condensing turbines. Since the inherent efficiency of industrial condensing steam turbines is on the order of 25 to 30%, the overall fuel utilization of the site falls off rapidly and ultimately approaches an asymptote in the range of 30%. Thus, processes with high ratios of power requirement to process heating load, such as steam crackers, are quickly forced into condensing

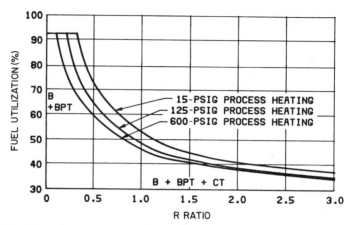

Fig. 6-8   Effect of steam pressure on fuel utilization for a steam-driver-only system. B, 1500-psig boiler; BPT, back-pressure steam turbine; CT, condensing steam turbine.

turbine drivers, and overall fuel utilization rapidly falls off. On the other hand, evaporation plants, distilleries, and other large users of process steam have very low R ratios and a high potential for fuel utilization, on the order of 70%.

Note that this type of analysis, although it does not use the available energy concept directly, is a direct practical application of second law thinking. As fuel utilization is increased, less fuel (with its attendant losses of available energy) is fired. Within the freedom allowed by other constraints, if one can reduce the overall ratio R for the site, the overall fuel utilization of the plant can be moved up the curve toward higher values.

As mentioned earlier, changes in the pressure at which the process steam is required will affect the best possible fuel efficiency in the system. Figure 6-8 shows the impact of several process steam temperatures on the best available fuel efficiency in a steam-driver-only case. Note that the impact is two-fold. The lower the process steam pressure, the higher the ratio R at which condensing turbines must be employed. So for a 15-psig system, back-pressure turbines will be in balance until the site ratio R exceeds about 0.3, as compared with about 0.2 for the 125-psig curve. Conversely, at 600 psig the ratio approaches 0.1 at this point. The overall efficiency curves are also higher for lower steam pressures, because a greater percentage of the power is being produced in the most efficient manner. Ultimately all these curves will approach approximately the same asymptote at very high values of R.

Obviously, the steam generation pressure will also affect the fuel utilization possible at any given ratio R. The curves presented so far have used a 1500-psig-fired boiler steam pressure. At lower pressures effects similar to the impact of process steam pressure will be seen. What controls the point at which condensing turbines must be employed is the work that can be extracted between the initial and final steam pressures in any turbine. These curves are based on a 75% turbine efficiency, but other curves can be calculated for any required machine efficiency. The following problem will allow the reader to demonstrate these effects for himself.

Problem: Maximum Potential Fuel Utilization

Calculate the maximum potential fuel utilization curve for a site that generates steam at 615 psig and 760°F and uses half the process heating steam at 15 psig and half at 235 psig. Assume turbine drivers operate at 75% efficiency.

Assume

Boiler thermal efficiency $= 0.9$

Turbine condensing pressure $= 1$ psia, $h = 70$ Btu/lb

Gross steam production $= 1.1 \times$ net production

Solution

Weighted average discharge $P = (15 + 235)12 = 125$ psig $= 140$ psia

From Molliere diagrams or steam tables, the unit enthalpy of steam entering the turbine is $h_{in} = 1385$ Btu/lb

From a Molliere diagram, $h_{out} = 1267$, so $w = 118$ Btu/lb steam flow

At 140 psia the latent heat $\Delta H_L = 942$ Btu/lb

At maximum utilization, $R = 118/942 = 0.125$

Fuel $= (1.1/0.9)[1385 - h_{L\,140}] = (1.1/0.9)(1060) = 1296$ Btu

$\eta = [(942 + 118)/1296] = 0.818$

At $R = 0.25$, $W = 236$ Btu/lb; 118 Btu will come from the back-pressure turbine, the rest must come from a condensing cycle

Power is derived from a condensing turbine at an exhaust pressure of 1.0 psia and at 75% efficiency $= 361$ Btu/lb steam (Molliere diagram)

Pounds of steam to condensing cycle $= (236 - 118)/361 = 0.327$ lb

Fuel $= (1.1/0.9)[1060 + (1385 - 70)0.327]$

$\qquad = 1.222[1060 + 1315(0.327)] = 1.222(1490) = 1820$

$\eta = [(236 + 942)/1820] = 0.647$

At other values of R, similar calculations give the results tabulated below.

R	W_{tot}	Condensing flow	Fuel utilization
0.25	236	0.327	0.647
0.5	472	0.980	0.518
0.75	608	1.357	0.473
1.0	942	2.283	0.407
1.5	1413	3.587	0.360

A. Gas Turbines and Diesel Drivers

Internal combustion drivers have significantly different ratios of kilowatts produced per unit of process heat load required for maximum fuel efficiency. For example, it was noted that a gas turbine required only about 1.0 Btu of process heat demand per Btu of power, whereas a 1500-psig back-pressure turbine exhausting to 125 psig required about 5 Btu of process heating load. Thus, it is possible to enhance the maximum potential fuel efficiency of complexes with high ratios of R by judicious driver selection. The impact of adding a gas turbine driver is shown in Fig. 6-9. The lower curve is the same one shown in Fig. 6-7 for a 1500-psig steam system with a process heating load at 125 psig. In the upper curve a combined-cycle gas turbine driver with an unfired waste heat boiler producing 600-psig steam that flows through a back-pressure turbine (as shown in Fig. 6-5) is used to replace part of the condensing turbine power requirement. The result is to essentially double the ratio R at which 70% fuel efficiency can be obtained, from 0.4 to about 0.8. Beyond $R = 0.8$, condensing turbine drivers are still required and the curve will flow down to an asymptote governed by this same low driver efficiency.

It is also possible to fire more fuel in the exhaust of the gas turbine, because its O_2 content is about 17%. This makes it possible to produce 1500-psig steam as opposed to 600-psig steam. The net effect is to increase still further the fuel utilization at ratios below about $R = 0.4$. But because this extra fuel is not used as efficiently as the first increment, the supplementary fired waste heat boiler efficiency curve falls below the unfired waste heat boiler curve at about a ratio of 0.4. It then takes a position about halfway between the steam turbine and unfired waste heat boiler curve as it flows out to high ratios of R. Thus, in the region of R between 0.2 and 0.4, a small increase in fuel efficiency can be achieved by supplementally firing the waste heat boiler on

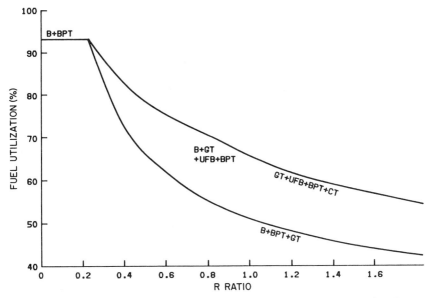

Fig. 6-9 Best achievable fuel utilization curves for steam power systems, based on a General Electric frame 5 gas turbine and 125-psig steam for process heating. B, 1500-psig, 955°F fired boiler; BPT, 75% efficiency back-pressure turbine; CT, 75% efficiency condensing turbine; GT, gas turbine; UFB, 630-psig, 755°F unfired waste heat boiler with gas turbine exhaust as the waste heat source.

the gas turbine. Above ratios of 0.4, however, this difference becomes negative.

In complex steam systems there are often several pressures at which process steam is used. Weighted average steam pressures can be used in creating the curves for your plant. The turbine efficiencies of small drivers may be markedly different from those of the main drivers, necessitating the use of weighted averages for turbine efficiency as well. This is not a trivial question, because often small single-stage turbines have efficiencies as low as 30%. The point is that the curves given here are typical and not directly applicable to a particular site (unless steam pressures and driver efficiencies happen to correspond). The curves can be calculated for each site from the turbine vendor's curves and steam tables in a straightforward manner.

In some installations diesel engines may be acceptable to the operating department as prime movers. This might be done directly or through a diesel-powered generating set. As noted earlier, diesel engines generate even more power per unit of process heat load required for maximum efficiency, because they use less excess air than a gas turbine. As shown in Fig. 6-10,

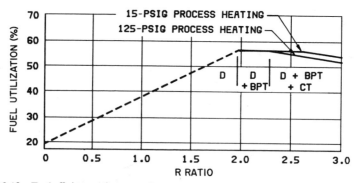

Fig. 6-10   Fuel efficiency of a generating system composed of a diesel engine plus an unfired waste heat boiler.

diesel engines can be employed to maintain the fuel utilization of plants with very high R ratios (above 2.0) at higher values than with steam turbines and gas turbine combinations alone. Thus, for very high R ratios, internal combustion engines need to be considered in maximizing fuel efficiency.

B. A General Relation for Fuel Utilization

Referring to Fig. 6-11, we can attempt to derive a generalized equation for the fuel utilization η as a function of R in terms of thermodynamic properties. We have shown that

$$R = \frac{\text{total shaft work load}}{\text{total process heat steam load}}.$$

In the general case,

$$R = (W_1 + W_2 + W_3 + W_4 + W_5)/Q_2 = \sum W/Q_2,$$

$$\eta = (W_1 + W_2 + W_3 + W_4 + W_5 + Q_2)/Q_1, \tag{6-2}$$

$$\eta = (Q_2R + Q_2)/Q_1 = Q_2(R + 1)/Q_1, \tag{6-3}$$

or

$$Q_1 = [Q_2(R + 1)]/\eta. \tag{6-4}$$

C. The Addition of a New Unit

Suppose an existing plant operates at $R = 0.5$ using 10,000 hp of steam turbine drivers and 125-psig steam for process heating. The power requirement corresponds to

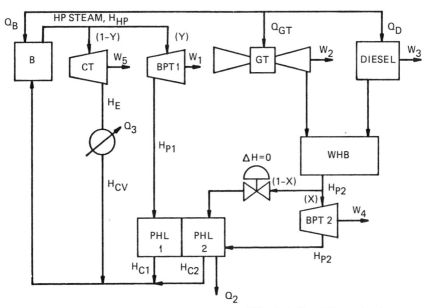

Fig. 6-11 Generalized steam power system. B, 1500-psig boiler; CT, condensing steam turbine; BPT, back-pressure steam turbine; PHL, process heat load; WHB, waste heat boiler; D, diesel engine; GT, gas turbine. Q = fuel = $\sum Q_i$; $Q_2 = \sum$ PHL.

$$10,000 \text{ hp} \times 2544 = 25,440,000 \text{ MBtu/h.}$$

The corresponding process steam load is $2(25.44) = 50.88$ MBtu/h.

From Fig. 6-7 we see that a steam turbine system requires some condensing turbine drivers and has an efficiency of 63%. The site fuel requirement is

$$Q_1 = (25.44 + 50.88)/(0.63) = 121.3 \text{ MBtu/h.}$$

Suppose it was planned to add a new plant requiring 50,000 hp plus 50 MBtu/h of process heat at 125 psig. The new R ratio would be $[25.44 + 5(25.44)]/(50.88 + 50) = 1.51$. The fuel utilization must come down to an overall value of 43% with steam drivers only, because most new machines must be condensing also (see Fig. 6-7):

$$\text{new fuel} = (152.8 + 101)/(0.43) = 590 \text{ MBtu/h,}$$

$$\text{incremental efficiency} = \Delta \text{ useful energy}/\Delta \text{ fuel}$$

$$= [50 + 5(25.4)]/(590 - 121) = 177/469 = 0.377.$$

If we chose a gas turbine to provide the new horsepower, the best overall efficiency possible would be -57% (see Fig. 6-9):

$$Q_1 = Q_2(R + 1)/\eta = (101)(2.51)/(0.57) = 445 \text{ MBtu/h},$$

$$\Delta \text{ useful energy}/\Delta \text{ fuel} = 177/(445 - 121) = 0.546.$$

The fuel savings would be $590 - 445 = 145$ MBtu/h.

By applying a factor for how much capital is justified for a given fuel savings, one can quickly decide whether gas turbine drivers should be considered further. If a 15% simple return is required and fuel costs $4/MBtu, then

$$\text{savings/capital} = 0.15.$$

For 8000-h/yr operation, $(8000 \times 145 \times 4)/(0.15) = \$30$ million is the capital justified.

This figure represents the *net* capital increase on the site which could be justified over the base cost of new facilities to achieve the indicated savings at a 15% simple return hurdle rate. The net capital increase includes not only the increase in the cost of gas turbine over steam turbine drivers, but also the decrease in base-case costs for systems to supply the extra fuel and reject the additional waste heat to the environment.

D. Motor Impact

In the case described previously, the new horsepower demand could be supplied by motors. The new R value would be

$$R = (25.44)/(50.88 + 50) = 0.25.$$

With steam drivers only (Fig. 6-7), the best possible fuel efficiency for the site would be about 85%. The new site fuel requirement would be

$$Q = (101 \times 1.25)/(0.85) = 148.5 \text{ MBtu/h}.$$

Incremental fuel would be $148.5 - 121 = 27.5$ MBtu/h plus the cost of 10,000 hp of electricity. The fuel requirement to supply 50 MBtu/h more 125-psig steam is about one-half the incremental duty, because the efficiency of the overall system is being increased.

To accomplish this enough extraction turbine capacity must be supplied (if not already available) in the existing drivers. This may mean an additional capital expenditure over the cost of the new motors and power supply. Whether this is an attractive case will largely depend on the cost of purchased power, but the thermodynamic efficiency of the complex will increase, because onsite condensing turbine power generation wll decrease.

The ratio approach thus provides a simple way to study options for new energy system loads. The possibilities and incentives can be quickly identified, giving direction to engineering studies.

E. Upgrading Existing Equipment

Start with the same plant as before, namely, 10,000 hp of steam turbine drivers, 125-psig steam for process heating, and an R ratio of 0.5. The incentives for switching some of the existing drivers to gas turbine drivers can be estimated from Fig. 6-9. Since no new loads are being added, R remains the same. By changing drivers we change the efficiency of the steam power system by eliminating condensing turbines. From the upper curve in Fig. 6-9, the maximum fuel utilization would be 77% versus 65% using steam turbines only. The new amount of fuel fired would be (from Eq. 6-4)

$$Q_1 = (51)(1.5)/(0.77) = 99.4 \text{ MBtu/h},$$

or a savings of about 22 MBtu/h. For a 15% simple return at a fuel price of $4/MBtu, the capital justified would be

$$\text{capital} = (8000 \times 22 \times 4)/(0.15) = \$4.7 \text{ million}.$$

Since we are talking about replacing *existing equipment,* there may be no credits for idling existing fuel supply or heat rejection equipment (i.e., the cooling tower) unless additional needs or system replacements are anticipated.

Another approach would be to use Fig. 6-8 to estimate the benefits for reducing the steam pressure required for process heating. If 15-psig steam could be used for all of the duty, the overall utilization at $R = 0.5$ could be 72% versus 65%. The new fuel consumption would be

$$Q = (51)(1.5)/(0.72) = 106.3,$$

a savings of 15 MBtu/h. Would the extra surface area be economic?

To summarize this section, the ratio R and appropriate fuel efficiency curves prepared for a site can be used to give a feel for the overall impact of changes in site demands and equipment. This can simplify efforts to understand site energy system interactions, save engineering work in planning improvements, and make sure some ideas are not overlooked during the brainstorming process.

IV. The Linear Programming Approach

A. Concepts

Linear programming is a mathematical technique often used to optimize the allocation of resources or the profitability of a complex sector of business. The technique requires that all possible options that may potentially be used be characterized mathematically to calculate their impact on an "objective function." This objective function could be fuel use, operating costs, net present value of the whole system, etc. The program calculates the objective function for all practical combinations of the potential options (as defined by system constraints) and selects the one that maximizes the desired objective.

The technique requires that options and constraints be represented (or approximated) by linear equations. By doing this very large numbers of simultaneous equations can be solved. Thousands of variables with upper and lower limits and hundreds of constraints (e.g., a steam balance) are generally feasible, and the optimum can be clearly identified.

Linear programming has its limitations. The solutions are derived for steady-state conditions only. Since nonlinear equations cannot be handled, the required linear approximation for any option may introduce misleading errors in some areas of the field to be studied. For example, capital costs are rarely linear with capacity; thus, sophisticated approximations are required. Yes/no decisions and discontinuous problems also require sophisticated treatment and/or off-line analysis.

Conceptually, a linear program is capable of selecting the optimum energy system in a complex case just as we have been doing by engineering judgment in the simple cases considered so far. Consider the simple array of possibilities shown in Fig. 6-12. Parcels of waste heat are available at A and B (sources), and heat is required at X and Y (sinks). Investments may be required to collect the heat, transport it, and use it at either X or Y or some combination of the two.

There are several constraints on this simple system, some of which are immutable (hard), and some of which may be changed with investment (soft). The hard constraints include the various heat and utility balances and production requirements. In addition, the engineer only poses options that are technically feasible. Other constraints include the capacities of the utility systems, energy transport systems, and individual pieces of equipment. Thus, only so much heat is available from A at temperature T_A for use at either X or Y. Recovery is also subject to the needs (balances) and temperature conditions at the user in question. The same is true for source B. Presumably, current consumption of most utilities would be reduced by this heat recovery effort, which would be backed through the energy supply systems and result in net savings in purchased available energy (fuel, power). The possibilities for using recovered energy in X and Y may vary continuously or in discrete steps.

Investments could be required at any point in the system. The impact of savings and costs on the objective function can be calculated for each

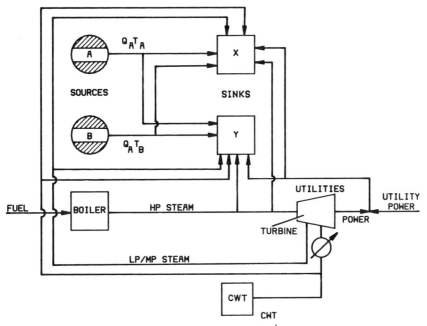

Fig. 6-12 Simple array at single capacity.

possibility identified by the energy conservation team. The program calculates all of the possible options and selects the one which maximizes the
objective function by a matrix algebra approach. This already seems formidable for the sample case, but often thousands of options must be explored
in a practical situation.

To accomplish this task the model requires a great deal of information
from the user. First, all possible options must be identified. This includes
"combination" or system options wherein several sources and sinks are tied
together in a system with appropriate backup components as described by
Manning.[1] The engineer must then quantify the energy saved, the type of
energy saved, the investment required, and any product value changes
which occur. This is normally done in a series of increments defined by the
practical constraints of the process units. An investment requirement is
estimated for each option, as is a downtime and an operating cost change
(other than energy). The information package for each idea is often termed a
"module." The program calculates the interactions of many modules on the
fuel and power requirements of the total system as inputs to the objective
function.

B. Practical Applications

Companies with reasonable backgrounds in linear programming for business planning can apply the approach to overall site energy surveys.[3] In one approach the net present value (NPV)[4] of utility savings less investment, operating costs, and expenses was set up as the objective function. Literally a thousand modules were considered in several time frames with different constraints. Comprehensive site energy plans were developed, which gave cohesiveness to individual project developments. The plan included the attractive technical opportunities plus constraints related to energy balance, capacities, financial considerations, and special site restrictions. The calculations treated the entire complex as an integrated system, evaluating interactions quantitatively. In addition, the potential for cooperative energy saving ventures with neighboring installations could be identified and pursued.

The specifics of the mathematical formulations will depend on the specific linear programming capacity at a given company. As a result, no specific procedures or examples will be given here. Operations research techniques are commonly applied, and provisions for future changes can be made. A number of major companies maintain a general system for linear programming applications.

Allan James[5] presented an example of overall site integration involving a petrochemical complex. Four processing units were integrated by a total energy system after intraunit energy integration opportunities were exhausted. The result is shown in Fig. 6-13. Over 9000 horsepower and 30 million kJ/h of waste heat were exchanged between the units in an economic way. The intent is to point out the magnitude of interactions developed in a single practical case through the use of a linear programming approach. Clearly, some or all of these steps might have been developed by an experienced engineering team working with the same data. However, an integration mechanism that depends on the mutual creativity of a number of lead process engineers is more likely to do a complete job when backed up by a systematic way of ensuring that the impact of all options is considered quantitatively.

Summary

This chapter covers a number of ways in which energy audit data can be used to improve overall site available energy use. These are treated in order of increasing complexity. First, several principles for onsite improvements are discussed, as shown in the following list.

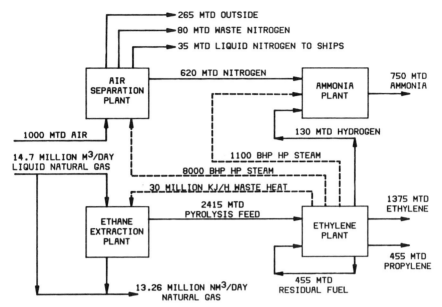

Fig. 6-13 An integrated petrochemical complex. MTD, metric tons per day.

(1) Ascertain whether all operating procedure improvements have been identified and implemented.
(2) Brainstorm for design improvements concentrating on:
     (a) systematically matching available energy supplies with process needs at corresponding levels,
     (b) reevaluating all of the assumptions inherent in the original design against current technical and economic constraints.

Overall site interactions are then considered. In simple cases the same procedures can be used and energy savings calculated back to their impact on fuel consumption. As complexity increases the ratio of total shaft work to total steam heat demand (R) can be used to indicate the potential best fuel utilization at the site, i.e., the maximum cogeneration potential, and the type of steam power system that gives the best efficiency. Internal combustion drivers are indicated at higher values of $R(>0.5)$.

Systematic ways of deriving the optimum combination of large numbers of ideas for site energy efficiency improvement generally involve linear programming. A simple description of the approach is given. Very complex arrays of possibilities can be handled by experienced practitioners of the art.

Notes to Chapter 6

1. E. Manning, Waste Heat Recovery Using a Circulating Heat Medium Loop. 45th American Petroleum Institute Refining Department Meeting, Houston, Texas, May, 1980.

2. W. O'Brien, Low Level Heat Recovery. Industrial Energy Conservation Technology Conference, Houston, Texas, April, 1982.

3. J. Guide and W. Lockett, Site Energy Surveys. Industrial Energy Conservation Technology Conference, Houston, Texas, April, 1981.

4. NPV is the current value of all past and future credits or payments discounted at the desired interest rates.

5. A. J. James, Design for Process Integration and Efficient Energy Utilization. Industrial Energy Conservation Technology Conference, Houston, Texas, April, 1982.

7

Methodology of Thermodynamic Analysis
General Considerations

Introduction

Thermodynamic analysis is a systematic way of characterizing what happens to the available energy of the materials in a process of interest. It is based on the usual heat and material balance for the process but extends it to encompass entropy changes and the attendant losses of available energy. Many articles, Ph.D. Theses, chapters in thermodynamic texts, and even entire books[1] have been written to describe the methodology of thermodynamic analysis. No attempt will be made to capture all of this material in this chapter. Indeed, our objective is to present a simplified approach adapted to an industrial environment. Some details that are of scientific importance but often not of practical significance will be omitted from the methods discussed here. Instead, the shortcuts learned from experience will be presented in an attempt to demonstrate that even a simplified analysis can assist in identifying fundamental process improvements. The following chapters will expand on the analysis of pervasive unit operations and then focus on using the data generated to find practical ideas for improving processes. Thus, the thrust of this section of the book will be a search for simple practical results as opposed to an erudite scientific analysis. The steps in the process are relatively simple at first glance. Very briefly, they are:

(1) obtain a consistent heat and material balance;
(2) determine the temperature of the surroundings, T_0;
(3) establish the boundary(ies) for the process segment(s) to be studied;
(4) compute the properties and heat effects for each critical stream location;
(5) compute minimum ΔA_i (ideal work) for the process and efficiencies;
(6) calculate A_{lost} (lost work) for each segment;

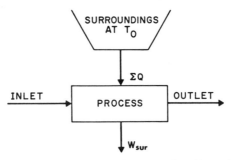

Fig. 7-1 General diagram of a process that exchanges work and heat with its surroundings.

(7) tabulate results in a way that characterizes the key elements of available energy consumption in the process to facilitate the study of improvements;

(8) break down sections in which large losses occur into more detail to pinpoint the step(s) that cause the losses.

Obviously, such a simple outline must be enhanced to be useful. A more detailed explanation of the procedure follows, using several examples to illustrate key points. The data generated in these examples will be used again in later chapters to generate ideas for process improvement.

Sign Conventions

Before proceeding with this discussion, sign conventions must be established. Consider Fig. 7-1. The essential features of a process that exchanges work and heat with the surroundings are shown. If heat transferred *into a process* is considered positive, then work transferred *out of a process* must also be positive. In effect, the process could convert a certain share of input heat to work or convert a share of input work to heat. Thus, the following sign conventions will be used:

(1) Q_{in} and W_{out} are positive;
(2) Q_{out} and W_{in} are negative.

I. Detailed Procedures

Consistent Heat and Material Balance Is Important

Before launching into the considerable arithmetic involved in thermodynamic analysis, it is important to be sure that the calculations are based on

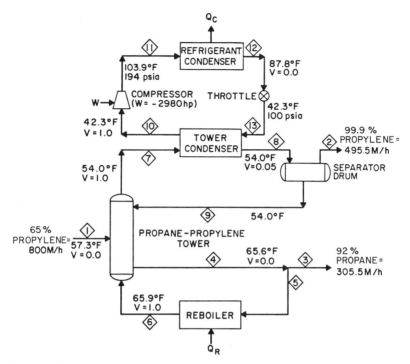

Fig. 7-2 Flow diagram for a propane–propylene separation process. V is the fraction of vapor present. From Reference 2.

an internally consistent foundation. The first step in determining the consistency of the data is to read the notes accompanying the tables used for enthalpy and entropy data (or the fine print in the computer program instruction manual). The reference states for all the data you are considering for a given material must be the same. *This is not trivial.* Even some data for steam have different reference states.

The data for CO_2 provide an example. In the third edition of Perry's "Chemical Engineers Handbook," the data for saturated CO_2 at 32°F are $s_{liq} = 1.0$ Btu/lb°R and $h = 36.7$ Btu/lb. In the tabulation for superheated CO_2 (from a different source), the 32°F data are $s_{liq} = 1$ and $h = 180$. As a result, to use the two tables together one must correct the saturated enthalpies by $180 - 36.7 = 143.3$ Btu/lb to make the data set consistent. The entropies, of course, need no correction.

Consider the propane–propylene separation process shown in Fig. 7-2, the thermodynamic data for which are shown in Table 7-1.[3] How does one determine whether the data are worthwhile?

A quick check of the overall material balance, including components, is

Table 7-1

Thermodynamic Data for a Propane–Propylene Separation Process[a]

Stream	Temperature (°F)	Pressure (psia)	Mole fraction vaporized	Flow rate (lb · mol/h)	Enthalpy (Btu/lb · mol)	Entropy (Btu/lb · mol°R)
①	57.28	120.0	0.0	800.0	−1455.102	18.306
②	53.99	120.0	1.0	494.45	4985.531	29.594
③	65.60	120.0	0.0	305.594	−1098.785	18.263
④	65.60	120.0	0.0	10,113.1	−1098.773	18.263
⑤	65.60	120.0	0.0	9,807.5	−1098.785	18.263
⑥	65.88	120.0	1.0	9,807.5	5560.5	30.934
⑦	53.99	120.0	1.0	9,889.0	4985.559	29.595
⑧	53.99	120.0	0.05	9,889.0	−1285.742	17.380
⑨	53.99	120.0	0.0	9,394.5	−1615.813	16.737
⑩	42.26	100.0	1.0	10,967.5	4873.434	29.673
⑪	103.90	194.0	1.0	10,967.5	5569.961	29.866
⑫	87.83	194.0	0.0	10,967.5	−664.8	18.473
⑬	42.26	100.0	0.116	10,967.5	−799.7	18.366

[a] Ambient temperature $T_0 = 536.67°R$ (77.00°F). From Reference 2.

the first step. For example, the propylene balance is

$$0.65(800) \stackrel{?}{=} 0.999(495.5) + (1 - 0.92)(305.5),$$

$$520 = 519.$$

The propane balance is

$$0.35(800) \stackrel{?}{=} 0.001(495.5) + 0.92(305.5),$$

$$280 = 0.5 + 281 = 281.5.$$

Considering all the significant figures listed in Table 7-1, this is not the best material balance, but it is acceptable.

Another initial step would be to check the overall property balance for the process and for individual pieces. The data in Table 7-1 already include composition effects. For example, around the tower condenser of Fig. 7-2, does $H_7 + H_3 = H_8 + H_{10}$?:

$$4985.6(9889) + 10,967.5(-799.7) \stackrel{?}{=} 9889(-1205.7) + 10,967.5(4873.4),$$

$$40,531,889 \neq 40,734,727,$$

$$\Delta = 203,000 \text{ Btu/h}.$$

This is very much less than the heat transferred in the condenser and can be neglected.

In another example, the enthalpy balance around the tower checks out. To check, calculate

$$H_1 + H_9 + H_6 \stackrel{?}{=} H_7 + H_4,$$

$$38,190,797 = 38,190,232.$$

Now consider the throttle valve in the refrigeration system. Theoretically (and in real life) throttle valves should be isenthalpic, i.e., constant enthalpy devices. The entropy should increase, but the total enthalpy in and out of the valve should be constant. As we see from streams 12 and 13, the data do not show this. Because of inaccuracies in the calculations, a decrease in enthalpy is shown. To avoid having this inconsistency mask the effects at the valve, an artificial correction is proposed. A loss of enthalpy corresponds to heat transferred out of the system. Therefore, the data can be corrected by postulating a heat leak at the valve as shown:

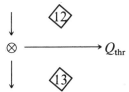

The magnitude of this correction is calculated from the following enthalpy change:

$$\Delta H = Q_{thr} = 10,967.5[-799.7 - (-669.8)],$$

$$Q_{thr} = -1.48 \times 10^6 \text{ Btu/h}.$$

According to convention Q_{thr} is negative because it flows out of the system. As long as we consider Q_{thr} in the calculations, this correction should not distort the process analysis.

For one last example of data verification, consider the compressor. The actual work done on the system is 2980 hp. This corresponds to 7.6×10^6 Btu/h. The enthalpy content of streams ⟨10⟩ and ⟨11⟩ corresponds to

$$W = \Delta H = 10{,}967.5(5569.96 - 4873.93),$$

$$W = \Delta H = 7{,}639{,}193.$$

The difference of 40,000 Btu/h is surely negligible.

By these checks we can verify the consistency of the enthalpy data by calculation. The entropy, of course, does *not* balance because in any real process entropy is created. However, the *direction* of entropy changes can quickly be verified. Is the overall ΔS positive? That is, is

$$S_2 + S_3 - S_1 \overset{?}{>} 0.$$

We obtain

$$494.5(29.594) + 305.6(18.26) - 800(18.306) > 0$$

by inspection.

In addition, entropy increases as temperature increases and streams are vaporized, i.e., S_6 is much bigger than S_5, $S_{10} > S_{13}$, etc.

A. Determining T_0 Is Often Arbitrary

The "dead-state" temperature is often taken as something approximating the average (or perhaps design) cooling water or air temperature. Sometimes there is an underground river or other relatively infinite heat sink to represent the surroundings.

In theory T_0 could be allowed to vary. This would greatly complicate the integration of

$$\int_1^2 T_0 \, dS. \tag{7-1}$$

In most practical cases this complication is not necessary. Sometimes summer and winter cases may be justified if wide changes in T_0 are experienced. In general, either the maximum or average T_0 is suitable.

On occasion a significantly higher value of T_0 may be in order. If a plant truly has no use for heat below 250°F, for example, engineers may wish to exclude considering the potential A between that temperature and a more conventional T_0. To focus the search for energy saving opportunities more sharply, T_0 could be taken as 250°F. This represents a form of prioritization, but could lead to ignoring a low-temperature opportunity.

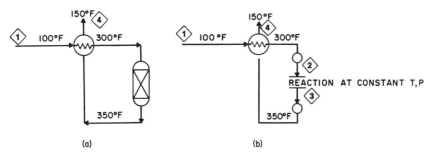

Fig. 7-3   A convenient boundary for reactions: (a), real process; (b), visualized process.

B. Boundaries and Reactions

In general, large process segments are considered first, with further break-down developed as areas of potential interest are identified. Thus, a stepwise procedure is established which goes into detail only for areas that need further breakdown to identify losses. For each segment, the process, feeds, products, heat and work effects, and utility usage must be determined. The portions of the process included (or excluded) from the segment analysis may have a large effect on the results and must be thought through. The same is true of utility systems. Does steam or electricity arrive across the boundary or are portions of the boilers or turbogenerator to be included? Is the refrigeration system in one segment or prorated over all segments using refrigeration? Generally, the available energy (work) gained by the cooling medium is counted as lost unless one purpose of the process is to heat water or other coolant.

It is sometimes convenient to treat reactions separately. There are two reasons for this: one, data are available only at a standard temperature; two, the lost work of the reaction may be ignored in the analysis because it is considered impossible to recover.

In the first approach the process is visualized with the reaction occurring at standard conditions (where data are available) and feed and products heated–cooled back to process conditions. This is diagrammed in Fig. 7-3, where the reaction is considerd to take place at the reference temperature and pressure of the available data. With this arrangement ΔH and ΔS can be obtained from standard data sources for many reactions. The book by Stull, Westrum, and Sinke[4] typically uses a standard reference state of the ideal gas at a pressure of 1 atm for each organic compound. Thus, the hypothetical reaction temperature should be chosen so that the feed stream and product stream are totally vaporized at a pressure of 1 atm. The assumption that the streams are ideal gases when totally vaporized at 1 atm is usually good.

The overall process entropy change ΔS and enthalpy change ΔH can be calculated as the sum of three parts. The entropy change $S_2 - S_1$ and enthalpy change $H_2 - H_1$ of the feed in going from the feed conditions to the hypothetical reaction conditions can be calculated by equations presented later in this chapter and in many textbooks.[5] The entropy change $S_3 - S_2$ and the enthalpy change $H_3 - H_2$ of the reaction must be calculated by hand. Be sure to include the entropy due to mixing effects in this hand calculation. The entropy change $S_4 - S_3$ and the enthalpy change $H_4 - H_3$ of the products going from the hypothetical reaction conditions to the product conditions are then calculated. The real entropy change of the process operation, $S_4 - S_1$, and the enthalpy change, $H_4 - H_1$, can be calculated by summing the three contributions:

$$\Delta S = S_4 - S_1 = (S_4 - S_3) + (S_3 - S_2) + (S_2 - S_1), \qquad (7\text{-}2)$$

$$\Delta H = H_4 - H_1 = (H_4 - H_3) + (H_3 - H_2) + (H_2 - H_1). \qquad (7\text{-}3)$$

C. Example

Analysis of the cracking of n-octane to 1-butene and n-butane illustrates how to calculate the entropy and enthalpy changes for a chemical reaction. The example illustrates the calculation techniques; it is not intended to represent an existing process. Consider the following reaction:

$$n\text{-octane} \longrightarrow 1\text{-butene} + n\text{-butane}.$$

The operating conditions for this process are illustrated in Fig. 7-4. Stull, Westrum, and Sinke[4] has been selected as the source of data on entropies and enthalpies of reaction. The energies of formation at 500 K and 1 atm for the compounds in this example are as follows:

Compound	Entropy (cal/g · mol K)	Enthalpy (kcal/g · mol)
n-octane	140.56	−55.50
1-butene	86.20	−2.70
n-butane	89.10	−33.51

Since the standard reference state for all three compounds is the ideal gas state at 1 atm, 1 atm is chosen as the hypothetical reaction pressure. The hypothetical reaction temperature is chosen as 500 K (440°F). At this temperature the feed and product streams will be totally vaporized.

After selecting hypothetical reaction conditions, the process operation can be modeled by the hypothetical process operation shown in Fig. 7-4.

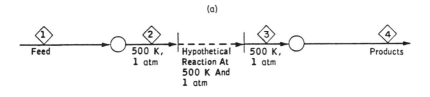

(a)

(b)

Fig. 7-4 The cracking of n-octane to 1-butene and n-butane: (a), real process operation; (b), hypothetical process operation.

From the reference data, the following values are calculated:

$$S_2 - S_1 = 0.101 \text{ kBtu/h°F},$$

$$H_2 - H_1 = -22.2 \text{ kBtu/h},$$

$$S_4 - S_3 = 0.148 \text{ kBtu/h°F},$$

$$H_4 - H_3 = 2.7 \text{ kBtu/h}.$$

The enthalpy change for a chemical reaction at constant temperature and with all the compounds at their standard reference states is

$$\Delta H_R = \sum n_i \Delta H_{f_i}^0, \tag{7-4}$$

where n_i is the net number of moles of compound i formed by the reaction. If moles of i are created by the reaction, n_i is positive; if moles of i are consumed by the reaction, n_i is negative. For the reaction under consideration,

$$\Delta H_R = H_3 - H_2 = (5.1 \text{ lb} \cdot \text{mol/h}) \Delta H_f^0 \text{ (1-butene)}$$

$$+ (5.1 \text{ lb} \cdot \text{mol/h}) \Delta H_f^0 \text{ (n-butane)}$$

$$- [(10.4 - 5.3) \text{ lb} \cdot \text{mol/h}] \Delta H_f^0 \text{ (n-octane)}$$

$$= 98.379(\text{kcal/g} \cdot \text{mol})(\text{lb} \cdot \text{mol/h}).$$

Correcting for units, we obtain

$$\Delta H_R = 177.1 \text{ kBtu/h}.$$

Entropy values for the pure compound in the standard reference state of the ideal gas at 1 atm, S_i^0, are tabulated in Stull, Westrum, and Sinke.[4] The entropy of any mixture in its standard state can then be calculated from the following equation:

$$S = \sum n_i S_i^0 + RN \ln N - R \sum (n_i \ln n_i), \qquad (7\text{-}5)$$

where n_i is the total moles of compound i in the mixture, N the total moles of all compounds in the mixture ($N = \sum n_i$), and R the gas constant. The last two terms in Eq. (7-5) account for the entropy of mixing effects. This is the entropy change upon mixing pure ideal gases to form an ideal gas mixture. For the two streams of interest,

$$S_2 = 1.462(\text{kcal/g} \cdot \text{mol K})(\text{lb} \cdot \text{mol/h}) = 1.462 \text{ kBtu/h}°\text{R}$$

$$- R(10.4) \ln(10.4)] \approx \text{lb} \cdot \text{mol/h}$$

$$= 1.462(\text{kcal/g} \cdot \text{mol K})(\text{lb} \cdot \text{mol/h}) = 1.462 \text{ kBtu/h}°\text{R}$$

and

$$S_3 = [(5.3)(140.56) + (5.1)(86.20) + (5.1)(89.10)$$

$$+ (1.987)(15.5) \ln(15.5) - (1.987)$$

$$+ (5.3 \ln 5.3 + 5.1 \ln 5.1 + 5.1 \ln 5.1)] \frac{\text{lb} \cdot \text{mol}}{\text{h}} \frac{\text{cal}}{\text{g} \cdot \text{mol K}}$$

$$= (1639.0 + 84.41 - 50.58) \text{ Btu/h}°\text{R} = 1.673 \text{ kBtu/h}°\text{R}.$$

Thus, $S_3 - S_2 = 0.211 \text{ kBtu/h}°\text{R}$.

D. Overall Entropy and Enthalpy Changes

The overall entropy and enthalpy changes can be calculated by summing the appropriate terms:

$$S_4 - S_1 = (S_4 - S_3) + (S_3 - S_2) + (S_2 - S_1)$$

$$= (0.148 + 0.211 + 0.101) \text{ kBtu/h}°\text{F} = 0.460 \text{ kBtu/h}°\text{F}$$

and

$$H_4 - H_1 = (H_4 - H_3) + (H_3 - H_2) + (H_2 - H_1)$$

$$= (2.7 + 177.1 - 22.2) \text{ kBtu/h} - 157.6 \text{ kBtu/h}.$$

The two quantities $S_4 - S_1$ and $H_4 - H_1$ are the thermodynamic quantities that are needed to proceed with a lost work analysis.

E. Another Approach

The standard free energy change for a reaction can also be used to calculate the entropy change if the enthalpy change is known. These calculations are often carried out at 77°F or other standard reference temperatures for the free energy data. The standard free energy change for the reaction at 77°F and 1 atm is

$$\Delta G = \Delta H_{537} - T_0 \, \Delta S_{537}, \tag{7-6}$$

$$\Delta H_{537} = \sum_i n_i \, \Delta H_{f_i}^0, \tag{7-7}$$

where $\Delta H_{f_i}^0$ is the standard heat of formation for the ith component and n_i the number of moles of that component.

With data on the free energy change for any reaction at 77°F and 1 atm pressure, the corresponding ΔS_{537} can be calculated from Eq. (7-6). The simplest way to obtain ΔG_{537} for any reaction is by the algebraic combination of the free energy changes of formation for the substances involved.[5] For example, for the reaction

$$C + \tfrac{1}{2}O_2 \rightarrow CO,$$

$$\Delta G_{537} \text{ (reaction)} = \Delta G_{537} \text{ formation of CO}), \tag{7-8}$$

since C and O_2 are elements and have zero ΔG_f^0.

Data on ΔG_{537}^0 (really $\Delta G_{T_0}^0$) do not exist for all reactions. For organic compounds they can sometimes be calculated from structural considerations. Sometimes direct data for ΔS_{537} or absolute data on the entropy of the products are available. For minor components the entropies of similar compounds may be assumed to apply.

The reactants and products must be returned to their real pressures and temperatures by hypothetical heat exchangers and engines. For temperature changes at constant pressure,

$$\Delta H = \sum n_i C_p \, \Delta T, \tag{7-9}$$

$$\Delta S = \sum n_i C_p \ln(T_2/T_0). \tag{7-10}$$

Remember that both enthalpy and entropy *increase on heating.*
For pressure changes and mixing,

$$\Delta S = \sum n_i R \ln(P_2/P_0). \tag{7-11}$$

Entropy always *increases when pressure drops.*

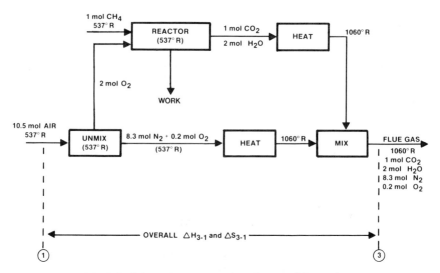

Fig. 7-5   Schematic representation of a reversible reaction.

For an ideal gas or a liquid, $dH/dP = 0$. For significantly nonideal fluids, activity (γ_i) or fugacity coefficients must be inserted for each component.

A diagrammatic summary of the approach for a reversible process is given in Fig. 7-5. The example shows the combustion of methane with 10% excess air producing flue gas at 600°F; this will be part of one of the examples discussed later in this chapter. The process assumes an isothermal reaction and reversible heat exchange and mixing. Since this reaction liberates available energy, the ideal work of the reaction represents the maximum that can be recovered.

Standard thermodynamic data are available for the reaction of CH_4 and O_2 (not air) at 77°F (537°R) and 1 atm. To use these data the air must first be separated in an "unmixing" step. An entropy change is associated with this step, in which 2 mol of O_2 are produced for the reaction at 537°R. The enthalpy and entropy changes in the reactor are calculated from the standard data. Both streams must then be heated reversibly to the outlet temperature (1060°R) and remixed to complete the process.

The overall ΔH and ΔS for the reactant side of the process are calculated by adding the individual changes for each step. The ideal work available from the process can be calculated from Eq. (7-21). These data are useful in evaluating real processes.

Of course, real processes do not operate reversibly and do not incorporate isothermal reactions. The usual route is represented in Fig. 7-6. The overall property changes ΔH_{3-1} and ΔS_{3-1} for the process will be the *same* for the

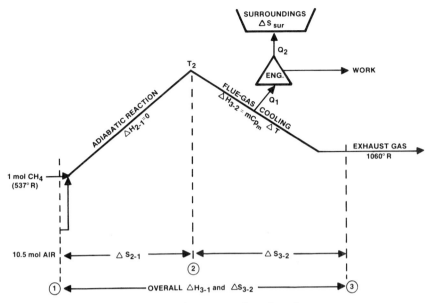

Fig. 7-6 Schematic representation of a real process.

real flue gas, because they have the same initial and final states. The difference will be in the amount of useful work that can be recovered.

The reaction is adiabatic in a real process. The reactants enter the reaction zone at T_0 (537°R) and are heated by the energy released in the reaction to T_2 (the adiabatic flame temperature). Calculation of T_2 is often done by trial and error because of the steep variation of the heat capacities of CO_2 and H_2O with temperature. The next step in a real process is to cool the flue gases (e.g., by generating steam), which is the means by which useful work (available energy) is recovered.

In most circumstances the ΔS of the cooling step, ΔS_{3-2}, can be calculated from the ideal gas equations given previously [see Eq. (7-10)]. Since the overall entropy change ΔS_{3-1} is known from the reversible analysis, ΔS_{2-1} (for the reaction) can be obtained by difference. The available energy loss (degradation) for the reactants in both the reaction and the cooling steps can be calculated from the entropy changes.

Note that the *total entropy change* for the cooling step must include the ΔS for the coolant as well as that for the flue gas. The ΔS described here covers only the flue-gas side. The hardware, the coolant, and the operating conditions of the coolant will have a major impact on the net available energy recovered.

F. Mixing

For ideal gases the entropy change of a pure substance mixed with other ideal gases can be calculated from its partial pressure P_i in the mixture in the same way as if it had expanded from the total pressure to its partial pressure:

$$\Delta S_{\mathrm{mix}} = R \sum n_i \ln(P/P_i).$$ (7-12)

For ideal solutions mole fractions X_i are substituted for partial pressures. There are no enthalpy or volume changes, by definition, so

$$T_0 \Delta S_{\mathrm{mix}} = \Delta A_{\mathrm{mix}} = T_0 R \sum X_i \ln X_i.$$ (7-13)

For nonideal solutions the term $\gamma_i X_i$ is substituted for X_i in the log term.

Of course, thermodynamic tables are available for many compounds and some mixtures (e.g., steam and air), obviating the need for some of these calculations. However, be sure to remember the earlier caveat to ensure the reference conditions of the data for each component are consistent.

G. Standard Chemical Availabilities

For some reactions data are available at standard temperature and pressure (STP), allowing calculation of the available energy changes of the reaction directly. Sussman[1] delineates the procedure for combustible chemicals, in which the standard free energy of formation G_f^0 is used to calculate the "standard chemical availability" of a fuel at STP according to

$$A^0(C_xH_yO_z) = xG_f^0[CO_2(g)] + \tfrac{1}{2}YG_f^0[H_2O(l)].$$

Again, to correct for data at other temperatures and pressures, the procedures outlined earlier must be used. Sussman provides data for a number of compounds at STP in his book.

H. Difficulties with Data

When dealing with materials for which data are not available from tables, there are a number of approaches that can be explored. Some are simple, some are complex. The task is to select a data method or source that gives consistent data over the range of interest. A number of thermodynamic texts treat calculation procedures. In addition to the book by Sussman[1] mentioned previously, there are two other data sources that have been found useful. These are Keenan, "The Thermodynamic Properties of Steam,"[6] and Stull, Westrum, and Sinke, "The Chemical Thermodynamics of Organic Compounds."[4]

For light hydrocarbons through heptane and nonhydrocarbon gases (for example, N_2), the Benedict–Webb–Rubin equation of state gives good, consistent results.

Linhoff[7] derives a simplified approach to obtaining entropy data from known data on enthalpy changes that minimizes errors through the use of ratios. He has found this to give good results for a number of cases. His method results in the following equation:

$$\frac{(S_2 - S_1)}{(H_2 - H_1)} = \left(\frac{1}{T_2 - T_1} \ln \frac{T_2}{T_1} \right). \tag{7-14}$$

With Eq. (7-14) the entropy change of a material going from state 1 to state 2 can be calculated from the known enthalpy change for the change in state. This can be extended to cover available energy changes:

$$\frac{A_2 - A_1}{H_2 - H_1} = \left(1 - \frac{T_0}{T_2 - T_1} \ln \frac{T_2}{T_1} \right). \tag{7-15}$$

As an example we can test the relation on some of the data in Table 7-1. Streams 4 and 6 are of the same composition, but one is vapor and one is liquid:

$$H_2 - H_1 = 5560.5 - (-1098.8) = 6659.3 \ \text{Btu/lb} \cdot \text{mol},$$

$$\frac{S_2 - S_1}{6659.3} = \frac{1}{T_2 - T_1} \ln \frac{T_2}{T_1} = \frac{1}{65.9 - 65.6} \ln \frac{525.9}{525.6}$$

$$= 160,005,706/0.3 = 0.001902,$$

$$\Delta S = 6659.3(0.001902) = 12.666 \ \text{Btu/mol°R}.$$

The data in Table 7-1 (calculated by complex computer correlations) give a value of

$$\Delta S = 30.934 - 18.263 = 12.671 \ \text{Btu/mol°R}.$$

This is an amazing agreement.

Turning to a less ideal substance, some conditions for superheated steam were chosen at random: 1500-psia steam at 950°F and 700°F. The data from the steam tables are shown in the following tabulation:

Parameter	700°F Steam	950°F Steam	Difference
h (Btu/lb)	1278.2	1460.1	181.9
s (Btu/lb°R)	1.4434	1.5791	0.1357

From Eq. (7-14) we have

$$\Delta s = [181.9/(950 - 700)] \ln(1410/1160) = 0.1420 \text{ Btu/lb}°\text{R}.$$

This also is a generally satisfactory agreement for initial calculations.

These examples were chosen at random from data readily at hand. They do not represent an extensive test of the approach, but along with Linhoff's examples represent some encouraging data. The reader should evaluate his own cases independently.

Parker[8] outlined a system for calculating available energy changes from enthalpy and density changes. Once pure components have been corrected to the stream compositions at T_0 and P_0 by Eqs. (7-12) an isothermal and then an isobaric correction is made to process conditions. Thus, data for H and S or A and γ_i are needed at atmospheric conditions for all components. The approach is as follows.

(1) Make an isothermal correction from (T_0, P_0) to (T_0, P) as follows:

$$\left(\frac{\partial A}{\partial P}\right)_T = V - (T - T_0)\left(\frac{\partial V}{\partial T}\right)_P. \qquad (7\text{-}16)$$

But $T = T_0$, so

$$\left(\frac{\partial A}{\partial P}\right)_T = V,$$

where V is the specific volume of the material.

(2) Make a constant pressure correction from (T_0, P) to (T, P):

$$\left(\frac{\partial A}{\partial T}\right)_P = \left(\frac{\partial H}{\partial T}\right)_P (1 - T_0/T). \qquad (7\text{-}17)$$

For single-phase systems $(\partial H/\partial T)_P$ is simply the heat capacity at constant pressure, C_p. However, when multiple phases are encountered, $(\partial H/\partial T)_P$ may include a vapor heat capacity, a latent heat of vaporization, and a liquid heat capacity. Numerical integration is often required with this approach, but results should be as consistent as the data at standard conditions.

Various other approaches are possible, such as the method of Grossman *et al.*[9] for petroleum fractions, but they will not be treated here. Try the simplest approach first, because the purpose of the analysis is to get *direction* for improvement, not precise scientific calculations.

II. Illustrative Examples

A. Calculation of Flows and Properties

After this excursion into handling data problems, let us turn to two examples and complete the preparation for the thermodynamic analysis by calculating flow rates, stream properties, work inputs, and heat duties at each critical location. The first breakdown of the process into segments generally flows naturally from this step.

1. A Propane–Propylene Splitter

For the propane–propylene splitter described in Fig. 7-2, most of the necessary data are shown in Table 7-1, but some quantities remain to be calculated. These consist primarily of the interactions with the surroundings Q_C, Q_R, and W. Both heat effects can be calculated from the enthalpy changes:

$$Q_C = H_{12} - H_{11} = -68.4 \text{ MBtu/h},$$

$$Q_R = H_6 - H_5 = 65.3 \text{ MBtu/h}.$$

The work (2980 hp) is merely a conversion of units,

$$W = -2980 \text{ hp} = -7.6 \text{ MBtu/h}.$$

The process can now be broken down into six segments for preliminary analysis. These are the tower, reboiler, tower condenser, compressor, refrigerant condenser, and throttle valve. The analysis of this process will be completed later in this chapter.

2. Steam Power Problem

For other processes completing the balances is not so trivial. Consider the steam power example shown in Fig. 7-7. One mole of methane and air at ambient conditions is to be burned to produce steam at 500 psia and 900°F, which in turn is to be used to generate power in a turbine. The thermodynamic data for the process are given in Table 7-2. Assume $T_0 = 77°F$.

Stoichiometry provides information about the flue-gas and steam flow rates. For 10% excess air the reaction for the combustion of CH_4 is

$$CH_4 + 2.20_2 + 8.3N_2 \rightarrow CO_2 + 2H_2O(g) + 8.3N_2 + 0.20_2, \tag{7.18}$$

producing 11.5 mol flue gas (with a heat capacity of 7.5 BTU/16 mol°F. The standard changes in enthalpy and entropy for the reaction at 77°F and 1 atm

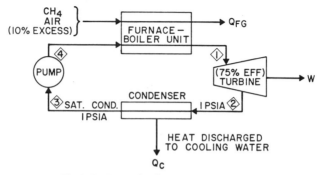

Fig. 7-7 Example of a steam power system.

can be calculated from tabulated data on the heats and free energies of formation for the reactants and products. Remember that for the elements these quantities are zero. The data[10,11] are shown in the following tabulation:

Compound	ΔH_f^0 (cal/g · mol)	ΔG_f^0 (cal/g · mol)
CH_4	−17,889	−12,140
CO_2	−94,050	−94,260
$H_2O(g)$	−57,798	−54,635

Adding the heats of formation, we obtain

$$H_R = -94,050 + 2(-57,798) - (-17,889) = 191,757 \text{ cal/g} \cdot \text{mol } CH_4,$$

or, in English units,

$$\Delta H_R = 345,163 \text{ Btu/lb} \cdot \text{mol } CH_4.$$

Similarly, ΔG_R is 191,390 cal/g · mol CH_4 = 344,502 Btu/lb · mol CH_4.

These quantities represent the maximum enthalpy and free energy available from the reaction. Also, remember that for an isothermal reaction the available energy and the free energy changes are equal, so that 344,502 Btu/lb · mol CH_4 also represents the maximum potential work that can be recovered from the reaction.

The reactants enter the system at T_0, but the reaction takes place at a higher temperature. This will change ΔH_R by the difference in heat capacities between products and reactants, as described by Obert,[12] but will not change the *maximum available energy* released in the reaction. The temperature and the process used to recover this energy do, however, have a marked effect on the net work derived from the potential 344,502 Btu/lb · mol CH_4 available.

Table 7-2

Thermodynamic Data for the Steam Power System of Fig. 7-7[a]

Stream	State of the steam	T (°F)	P (psia)	H (Btu/lb)	S (Btu/lb°R)
①	Superheated vapor	900	500	1465.1	1.6975
②	Wet vapor, $X - 0.973$	101.74	1	1077.5	1.9275
③	Saturated liquid	101.74	1	69.7	0.1326
④	Subcooled liquid	101.74	500	69.7	0.1326

[a] From Reference 2.

By definition $\Delta G_{537} = \Delta H_{537} - T_0 \Delta S_{537}$ at 1 atm and constant temperature, so that ΔS_{537} for the reaction can be calculated from

$$\Delta S_{537} = -[(\Delta G - \Delta H)/T_0] \quad \text{or} \quad [(\Delta H - \Delta G/T_0] \qquad (7\text{-}19)$$
$$= [(345{,}163 - 344{,}502)/537] = 1.231 \text{ Btu/lb} \cdot \text{mol°R}.$$

This represents the minimum ΔS for the process. Thus, the property changes for the reaction at T_0 and 1 atm are available. Enthalpy balances allow calculation of the remaining quantities.

Discarding the flue gas at 600°F corresponds to a heat effect with the surroundings. The heat effect Q_{FG} can be calculated from the data on p. 155:

$$Q_{FG} = 11.5 \text{ mol} \times 7.5(77 - 600) = 45{,}109 \text{ Btu/mol}.$$

The heat available for producing steam is, by difference, $\Delta H_R - Q_{FG}$, assuming no losses from the furnace. Obviously, any losses could be subtracted as a further correction:

$$Q_{stm} = 345{,}163 - 45{,}109 = 300{,}054 \text{ Btu/mol CH}_4.$$

The amount of steam produced is

$$M_{stm} = (300{,}054)/(h_1 - h_3) = (300{,}054)/(1465.1 - 69.7)$$
$$= 215 \text{ lb/mol CH}_4.$$

The amount of work produced by the turbine is

$$W = 215(h_1 - h_2) = 83{,}334 \text{ Btu/mol CH}_4.$$

Since this work is exported from the system, it is positive.

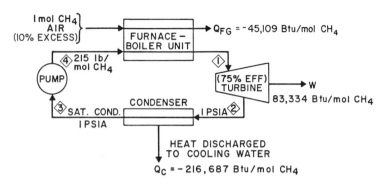

Fig. 7-8 Completed heat and material balance for the steam power system example.

The last quantity, Q_C, is similarly calculated from the steam flow and enthalpy values:

$$Q_C = 215(h_3 - h_2) = -216,687 \text{ Btu/mol C.}$$

The results are shown in Fig. 7-8.

This problem demonstrated in more detail the calculations needed to complete the mass and first law energy balances to establish the basis for thermodynamic analysis. When tabulated data are available, the property changes for the reaction can be obtained directly at the standard temperature. This process can also be broken down into segments for preliminary analysis as follows: the furnace–boiler, turbine, condenser, and pump.

We are now ready to complete the thermodynamic analysis of both processes.

B. Calculating the ΔA_i (Ideal Work)

The ΔA_i (ideal work) of any process is the best that can be done with the process, i.e., either the minimum required input or the maximum available energy recovery for the essential element of the change being achieved. The purpose of making the calculation is to establish how far the proposed process is from the ideal. For the propane–propylene separation process, the essential elements are to produce 494.5 mol/h of 99.9% pure propylene and 305.5 mol/h of 92.8% pure propane from 800 mol/h of 65% propylene feed. For the steam power problem, the essential elements are to burn fuel and produce power.

For any process the ideal ΔA is calculated only from the initial and final states of the materials involved. Thus, for the separation process only flow rates, compositions, and conditions of the feed and products enter into the calculation:

$$\Delta A_i = \Delta H - T_0 \Delta S$$
$$= (H_2 + H_3 - H_1) - T_0(S_2 + S_3 - S_1). \qquad (7\text{-}20)$$

From Table 7-1,

$$H_1 = (800)(-1455.102) = -1,164,081.6,$$
$$H_2 = (494.45)(4985.531) = 2,465,095.8,$$
$$H_3 = (305.594)(-1098.785) = -335,782.1,$$
$$\Delta H = H_2 + H_3 - H_1 = 3,293,394.8,$$
$$S_1 = (800)(18.306) = 14,644.8,$$
$$S_2 = (494.45)(29.594) = 14,632.7,$$
$$S_3 = (305.594)(18.263) = 5581.1,$$
$$\Delta S = S_2 + S_3 - S_1 = 5568.96,$$
$$T_0 \Delta S = (536.67)(5568.96) = 2,988,695.5,$$
$$W_i = 3,293,394.8 - 2,988,695.5 = 0.3047 \times 10^6.$$

The actual work expended on the process is done at the reboiler and the compressor. However, in this case heat is supplied at T_0, already at equilibrium with the surroundings, and the only work expended is at the compressor, 7.6 MBtu/h. Thus, the thermodynamic efficiency of the process is $W_i/W_{acf} = 0.3/0.7 \approx 4\%$.

As a contrast, we calculate the ideal work for the compressor alone:

$$W_i = T_0 \Delta S - \Delta H \qquad (7\text{-}21)$$
$$= T_0(S_{11} - S_{10}) - (H_{11} - H_{10})$$
$$= T_0(10,967.5)(29.866 - 29.673)$$
$$- 10,967.5(5569.96 - 4873.43)$$
$$= -6.5 \times 10^6 \text{ Btu/h},$$
$$W_{act} = 2980 \text{ hp} = -7.6 \times 10^6 \text{ Btu/h},$$
$$\text{efficiency} = 85.5\%.$$

The efficiencies serve only to flag areas of the process that may be good candidates for improvement. There is much room for improvement in the process itself, but the machinery section is not likely to be fertile ground.

In the steam power example the ideal work is controlled by the combustion process. We can do no better than to recover all of the available energy liberated in this reaction as work at the turbine. In an ideal process the flue gases would be cooled to T_0 before leaving the system, and the reaction would take place in a reversible fuel cell as discussed in Chapter 2. The entropy change for the reaction is 1.231 Btu/mol CH_4 °R at 77°F and 1 atm, as calculated from the free energy of formation of the reaction in the previous section. The reactants enter the system not far from these conditions in the real world as well:

$$\Delta A_i = T_0 \, \Delta S - \Delta H = 344{,}502 \text{ Btu/mol } CH_4 \text{ burned.} \qquad (7\text{-}22)$$

Since 83,344 Btu were recovered as work, the efficiency of the process is about 24%. This is better than the separation process, but no bargain.

C. Calculate Available Energy Losses

At this point the overall efficiency of each process is known relative to its thermodynamic "best." The next step is to identify those segments of the process in which available energy (work) is lost (degraded) and to quantify each loss. The total of these losses and the ideal available energy change (work) should add up to the actual work (available energy) required or recovered from the process:

$$W_{act} = W_i - W_{lost}. \qquad (7\text{-}23)$$

Remember that W_{lost} is always a positive quantity. In a *work-consuming* process, W_i is *negative* and W_{act} is greater than W_i. For a *work-producing* process, W_i is *positive* and W_{act} is less than W_i. The final step in the methodology is really a series of steps. The two processes have already been broken down into several segments for analysis. The methods outlined in this chapter and in Chapter 2 will now be used to demonstrate how the analysis is performed.

First, the minimum availability and the actual availability changes are calculated for the whole process. The difference between these two quantities gives the lost work for the whole process. Then the process is decomposed into subprocesses and the availability changes and the lost work are calculated for each subprocess. Decomposition into smaller subprocesses can be done a number of times. Decomposition is stopped, and the second law analysis is considered to be complete, when every major contributor to the lost work has been identified. At this stage some subprocesses will be individual pieces of equipment, such as a heat exchanger, whereas others may be groups of equipment, such as a distillation tower and its reboiler and condenser. The results will be tabulated for use in Chapter 9, where implementation of the thermodynamic analysis for process improvement is discussed. These examples will be presented in terms of "work," but remember that available energy is equivalent to work in this process.

1. Propane–Propylene Example

From previous sections the lost work for process streams can be calculated from

$$W_{lost} = T_0\,\Delta S - Q. \tag{7-24}$$

For the reboiler,

$$W_{lost} = T_0(S_6 - S_6) - Q_R,$$
$$Q_R = 65.3 \times 10^6 \text{ Btu/h}.$$

Note that Q_R is taken from the surroundings at 77°F, or T_0, so that no available energy is input to the process at the reboiler. (More will be said about this in Chapter 8.) We therefore have

$$W_{lost} = 536.7(9807.5)(30.934 - 18.263) - 65.3 \times 10^6 = 1.4 \times 10^6 \text{ Btu/h}.$$

For the throttle valve,

$$W_{lost} = T_0(S_{13} - S_{12}) - Q_{thr} \quad \text{(remember } Q_{thr}\text{?)}$$
$$= 536.7(10{,}967.5)(18.366 - 18.473) + 1.48 \times 10^6$$
$$= 0.9 \times 10^6 \text{ Btu/h}.$$

For the refrigeration condenser,

$$W_{lost} = T_0(S_{12} - S_{11}) - Q_C,$$
$$Q_C = H_{12} - H_{11} = -68.379,$$
$$W_{lost} = T_0(S_{12} - S_{11}) - (-68.379)$$
$$= (536.7)(18.473 - 29.866)(10{,}967.5) + 68{,}380{,}000$$
$$= -67{,}062{,}128 + 68{,}380{,}000 = 1.3 \times 10^6 \text{ Btu/h}.$$

For the compressor,

$$W_{lost} = W - W_i = (7.6 - 6.5) = 1.1 \times 10^6 \text{ Btu/h}.$$

For the tower itself,

$$W_{lost} = T_0[S_7 + S_4 - (S_1 + S_6 + S_9)] - Q, \qquad Q = 0$$
$$= 536.7(477{,}358.7 - 475{,}269) = 1.12 \times 10^6 \text{ Btu/h}.$$

Table 7-3

Lost Work Results for C_3 Separation[a]

	Work (MBtu/h)			
Process step	Actual work	Ideal work	Lost work	Percentage of total lost work
Tower			1.1	14.7
Reboiler			1.4	18.6
Tower condenser			1.7	22.7
Compressor	7.6	6.5	1.1	14.7
Refrigerant condenser			1.3	17.3
Throttle			0.9	12.0
Entire process	7.6	0.3	7.5	100
		7.8		

[a] From Reference 2.

For the tower condenser,

$$W_{\text{lost}} = T_0(S_{10} + S_8 - S_7 + S_{13}) - Q,$$

where $Q = 0$ because losses are ignored and the entropy change of all streams is included; therefore,

$$W_{\text{lost}} = 1.7 \times 10^6 \text{ Btu/h.}$$

Various combinations of the same simple equations have been used to analyze this process. The results are tabulated in Table 7-3, where only absolute values are presented. The total lost work is approximately equal to the actual work required by the process. If this were not the case, significant errors would exist. This represents a final check on the consistency of the analysis. Note that all segments of the process contribute comparable shares of the lost work of the whole.

2. Steam Power Example

As indicated earlier, the following segments of this process will be explored: the furnace boiler, the turbine, the pump, and the condenser. Since work requirements are small at the pump, it will be ignored.

For the furnace boiler both sides of the system must be considered to account for the total lost work of the combustion and recovery process, just as they were for the tower condenser in the previous example. The work lost because the flue gas is not cooled back to T_0 is also included here. We have

$$\Delta S = M_{\text{stm}}(S_1 - S_4) + \Delta S_R,$$

Table 7-4

Actual and Lost Work for Steam Power System[a]

Parameter	Work (Btu/lb · mol C)	Percentage of ideal work	Percentage of total lost work
Actual work produced	83,334	24.0	—
Work lost from furnace and boiler	226,446	65.6	86.2
Work lost from turbine	26,555	7.7	10.2
Work lost from condenser	9,457	2.7	3.6
Work lost from pump	Neglected	—	—
Total lost work	261,135	100	100
Total lost work plus actual work	344,469		

[a] From Reference 2.

$$W_{lost} = T_0\, \Delta S - Q_{FG} = 537[(215)(1.6975 - 0.1326) - 1.231] + 45.109$$
$$= 225,123 \text{ Btu/mol CH}_4.$$

Thus, most of the loss in the available energy of the fuel occurs in this segment, as anticipated from Chapter 2. For the turbine,

$$W_{lost} = T_0 M_{stm}(S_2 - S_1) - Q.$$

Since no heat is transferred to the surroundings at the turbine (losses are neglected), the equation for W_{lost} becomes

$$W_{lost} = (537)(215)(1.9275 - 1.6975) = 26,555 \text{ Btu/mol CH}_4.$$

For the condenser,

$$W_{lost} = T_0 M_s(S_3 - S_2) - Q_C$$
$$= (537)(215)(0.1326 - 1.9275) - (-216,687) = 9457 \text{ Btu/h}.$$

In this case work is produced and the actual work is less than the ideal work. Hence, the ideal work equals the sum of the actual work and several lost work terms.

The energy available for the ideal process was calculated to be 344,502 Btu/mol CH$_4$. This checks reasonably well with the total of actual and "lost" work of 344,469 shown in Table 7-4. Losses are shown as a share of both the "ideal" work of the process and the "lost" work. Note that the vast majority of the losses occur in the furnace, as would have been predicted from earlier discussion.

In Chapter 9 the results from both these problems will be analyzed for energy efficiency improvements.

Summary

In this chapter the step-by-step methodology for thermodynamic analysis of a process is demonstrated primarily by working through two different examples to tabulations of the lost work (available energy) for simple segments of each process. Evaluation of the data and corrections for obviously wrong data are discussed. Methods of approximating missing data on entropy are provided along with several useful references.

The simplified approach demonstrated can provide insight into many processes. The objective is to identify the segments of each process in which the largest inefficiencies occur so that creative efforts can be focused there. Note that precise scientific analysis is far less important than are general directions to the practical energy conservation engineer.

Notes to Chapter 7

1. M. Sussman, "Availability (Exergy) Analysis." Mulliken House, Lexington, Massachusetts.

2. W. F. Kenney and R. Rathore, "Thermodynamic Analysis For Improved Energy Efficiency. AIChE Continuing Education Series, New York, 1980.

3. D. Maloney and L. Domash, "Analysis of Processes by the Second Law of Thermodynamics." Exxon Research and Engineering Company Report No. EE-57E-78, 1978. Negative values occur because the basis for the data is 77°F.

4. O. R. Stull, E. F. Westrum, and G. C. Sinke, "The Chemical Thermodynamics of Organic Compounds." Wiley, New York, 1969.

5. B. F. Dodge, "Chemical Engineering Thermodynamics," p. 498. McGraw-Hill, New York, 1944.

6. J. H. Keenan and F. G. Keyes, "The Thermodynamic Properties of Steam." Wiley, New York, 1936.

7. B. Linhoff, A Thermodynamic Approach to Practical Process Network Design. Ph.D. Thesis, Department of Chemical Engineering, University of Leeds, Leeds, England.

8. A. L. Parker, Available Energy Calculations for Process Engineers. Paper presented at the Industrial Energy Conservation Technology Conference, Houston, Texas, April, 1982.

9. E. D. Grossman, S. V. Smith, and J. C. Sweeney, Calculation of the Availability of Petroleum Fraction. Department of Chemical Engineering, Drexel University, Philadelphia, Pennsylvania.

10. Adapted from H. Van Ness, Thermodynamics for process evaluation. *Petroleum Refiner* **35**, 165 (1956).

11. J. Perry, ed., "Chemical Engineers Handbook," 3rd ed. McGraw-Hill, New York, 1950.

12. H. Obert, "Thermodynamics," p. 379. McGraw-Hill, New York, 1938.

8

Detailed Thermodynamic Analysis of Common Unit Operations

Introduction

This chapter will go into the thermodynamic analysis of certain pervasive unit operations in more detail. This is basically an extension of Chapter 7, "Methods of Thermodynamic Analysis: General Considerations," to provide a more thorough understanding of where losses occur in the process steps commonly used. We shall also be developing a basis for guidelines to minimize lost work in design and process development.

There is no guarantee that this chapter will treat all of the unit operations in which the reader might be interested. The scope of the discussion will be limited by the experience and research interests of the author.

I. Heat Exchange

A. General Analysis

In general, heat exchangers can be represented by the accompanying diagram. The directions of the flow streams or the numbers of passes do not affect thermodynamic analysis, because properties depend only on initial and final states.

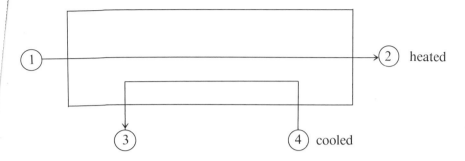

The lost work W_{lost} for any process step was derived in Chapter 2 as follows:

$$W_{lost} = T_0 \, \Delta S_{proc} - Q = -\Delta A_{tot}$$
$$= (A_2 + A_4) - (A_1 + A_3) = (A_3 - A_4) + (A_1 - A_2)$$
$$= (H_3 - H_4) + (H_1 - H_2) - T_0[(S_3 + S_1) - (S_4 + S_2)]. \quad (8\text{-}1)$$

Ignoring losses, $(H_1 - H_2) = -(H_3 - H_4)$ and the enthalpy terms sum to zero; therefore

$$W_{lost} = - T_0(-\Delta S_{tot}) = T_0 \, \Delta S_{tot}. \quad (8\text{-}2)$$

If only the process side properties are known, Q is calculated as follows:

$$Q = \Delta H_{proc} = -\Delta H_{util}. \quad (8\text{-}3)$$

B. Impact of Heat Source

Consider the reboiler in the fractionation problem of Fig. 7-2, in which 65.3 MBtu/h are supplied at 77°F to reboil the tower. An entropy change occurs on the process side, but because the heat is supplied to T_0, Q can be used directly in calculating W_{lost}:

$$W_{lost} = T_0(S_6 - S_5) - (H_6 - H_5)$$
$$= (536.7)(9807.5)(30.934 - 18.263) - (9807.5)[5560.5 - (-1098.8)]$$
$$= 1.4 \text{ MBtu/h}.$$

Is this insensitive to the source of heat? That is, is it always equal to 1.4 MBtu/h regardless of the source of heat?

1. Assume 0-psig Steam is the Heating Medium

The lost work is calculated from Eq. (8.2) as follows:

$$W_{lost} = T_0[\Delta S_{proc} + S_v - S_l],$$

where S_v and S_l are the entropy of the vapor and liquid, respectively. The steam flow rate is

$$M = (65.3 \text{ MBtu/h})\Delta H_v = (65.3 \times 10^6)/(970.3) = 6.7299 \times 10^4 \text{ pph}.$$

The change in entropy[1] is

$$S_v = 1.7566 \text{ Btu/lb}°\text{R},$$

$$S_l = 0.312.$$

The lost work is calculated as follows:

$$\begin{aligned}
W_{lost} &= T_0[S_6 + S_1 - S_5 - S_v] \\
&= (536.7)[(9807.5)(30.934) + (6.7299 \times 10^4)(0.3120) \\
&\quad - (9807.5)(18.263) - (6.7299 \times 10^4)(1.7566)] \\
&= (536.7)(303,385.2 + 20,997.3 - 179,114.4 - 118,217.4) \\
&= (536.7)(27,050.7) = 14,518,110.7 \text{ Btu/h} = 14.52 \text{ MBtu/h}.
\end{aligned}$$

This is 10 times the magnitude calculated previously, indicating a large dependence on the temperature (available energy level) of the heating medium.

2. Assume Steam at 77°F is the Heating Medium

In the original problem it was assumed that heat was available from the surroundings at $T_0 = 77°$F. Let us consider supplying heat from steam saturated at 77°F and calculate the work lost from the total entropy change.

At 77°F the latent heat of vaporization is 1050.4 Btu/lb. The steam flow rate is

$$M = (65,300/1050.4) = 62,166.8 \text{ pph},$$

and the rest of the calculation proceeds as follows:

$$S_1 = 0.0876 \text{ Btu/lb}°\text{R}, \qquad S_v = 2.0445 \text{ Btu/lb}°\text{R},$$

$$W_{\text{lost}} = T_0 \, \Delta S_{\text{tot}} = T_0[(9807.5)(30.934) + (62{,}166.8)(0.0876)$$

$$- (9807.5)(18.263) - (62{,}166.8)(2.0445)]$$

$$= T_0[308{,}831 - 306{,}214.4],$$

$$T_0 = 536.7,$$

$$W_{\text{lost}} = (536.7)(2616.6) = 1{,}404{,}329 \text{ Btu/h}.$$

Thus, the temperature level of the heating medium is the critical property that affects the lost work in a heat exchanger. Relating this to the concept of reversibility, if heat crosses the process boundary at T_0, the exchange with the environment is essentially reversible. Work is still lost in transferring heat to the process because a finite temperature difference exists. Heating media at temperatures T_h above T_0 can be thought of as undergoing two phases of lost work: one from T_h to T_0 and the 1.4 MBtu/h already calculated from T_0 to the process temperature.

C. Lost Work Calculations Without Entropy

The same results can be approximated from knowledge of T and Q only. For an exchanger transferring heat above T_0 at constant (or average) temperature T, the reversible work obtainable (or required) from Q is

$$W_R = Q(1 - T_0/T).$$

Referring to the diagram given earlier in this section, the work available from the cooled stream at mean temperature T_h would be

$$W_{\text{out}} = Q(1 - T_0/T_h),$$

and the work required to pump the same amount of heat from T_0 to the mean temperature of the heated stream (T_c) would be

$$W_{\text{in}} = Q(1 - T_0/T_c).$$

The work that might be accomplished by the exchanger is the difference between the two:

$$W = Q[(1 - T_0/T_h) - (1 - T_0/T_c)] = T_0 Q[(T_h - T_c)/T_h T_c].$$

As we shall see later, the same equations basically apply to refrigerated exchangers as well.

Because no work is obtained from the exchanger, *the work potentially available is, in fact, the lost work of the exchanger.* To demonstrate this, let us recalculate the two reboiler cases using temperatures and Q only. The mean temperature of the process stream (from Fig. 7-2) is 65.75°F (525.75°R); $Q = 65.3 \times 10^6$ Btu/h.

(1) *For 77°F heat,*

$$W_{lost} = \frac{(537)(65.3 \times 10^6)(537 - 525.75)}{(537)(525.75)} = 1.397 \times 10^6 \text{ Btu/h.}$$

(2) *For 0-psig steam,*

$$W_{lost} = \frac{(537)(65.3 \times 10^6)(672 - 525.75)}{(672)(525.75)} = 14.516 \times 10^6 \text{ Btu/h.}$$

These results obviously agree very well with those calculated from stream properties.

D. Lost Work Calculated from the Total Available Energy Change in the Exchanger

In an exchanger changes in available energy are of opposite sign: the heating medium gives up available energy, whereas the heated stream gains available energy. The difference between the two changes, i.e., the algebraic sum of the total change in available energy, equals the negative of the iost work for the exchanger:

$$- W_{lost} = \Delta A_h + \Delta A_c = \Delta H_h - T_0 \Delta S_h + \Delta H_c - T_0 \Delta S_c. \quad (8\text{-}4)$$

The algebraic sum of enthalphy changes must be zero if losses are ignored. Thus,

$$- W_{lost} = T_0(\Delta S_h + \Delta S_c). \quad (8\text{-}5)$$

Considering the reboiler discussed previously,

$$- W_{lost} = - T_0[(62,166.8)(0.0876 - 2.0445) + 303,385.2 - 179,114.4]$$

$$= - T_0(-121,654.2 + 124,270.8)$$

$$= (-536.7)(2616.6) = 1,404,329.2 \text{ Btu/h.}$$

Consequently, the lost work of any exchanger can be calculated from the total available energy changes of the streams if it is more convenient.

Figure 8-1 reiterates the conclusion of the foregoing development graphically: smaller driving forces result in less entropy production and, therefore, less work is lost in heat exchange and also in other unit operations. The graph shows qualitatively the comparison between two heat exchangers

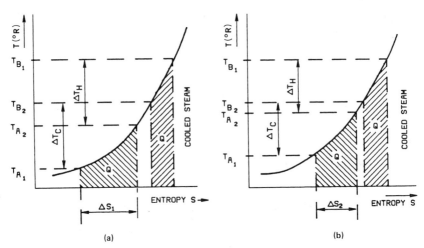

Fig. 8-1   Temperature versus entropy for two heat exchangers performing the same task at constant pressure. Graph (b) has a smaller mean temperature difference than graph (a), i.e., $\Delta T_{(a)} > \Delta T_{(b)}$ and $\Delta S_1 > \Delta S_2$.

doing the same job at constant pressure on a temperature–entropy diagram. One uses a smaller mean temperature difference than the other (and therefore must have either more surface or more pressure drop to improve its coefficient). Because the job is the same, the curves and the areas shaded are the same in both graphs, but the heated fluid area is moved closer to the area of the cooled fluid in part (b). This results in a smaller temperature difference for the exchanger and, graphically, a taller shape for the area Q, with a resultant smaller horizontal dimension ΔS. These relationships hold for any exchanger and for other processes as well.

In a report for the U.S. Department of Energy,[3] the relationship between lost work and heat transfer area was calculated for a number of cases. A typical curve is shown in Fig. 8-2, where an effectiveness ratio (lost work/ heat transfer area) is plotted against the total heat transfer area. The ratio (and the lost work) falls off very rapidly as ΔT decreases (area increases) until it begins to approach an asymptote at an area of 1000 Btu/h·ft² at 6000 ft². Increases in area beyond about 6000 ft² have little benefit to efficiency and are likely to be uneconomic for the case shown.

Similar relationships were demonstrated in Table 3-3 in terms of simple return on investment for exchanger area alone. As the log mean temperature difference (LMTD) decreased, so did return. As will be discussed in Chapter 10, the return on any efficiency project must include the investment impact on the utility system as well to give a true picture of the economics.

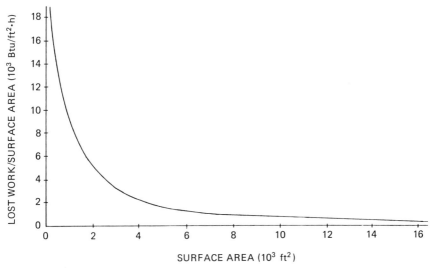

Fig. 8-2 Typical trade-off of effectiveness versus heat exchanger surface area.

However, the shape of the curve in Fig. 8-2 clearly indicates that there is a point beyond which further efficiency cannot be justified.

E. Insulation Loss

Heat loss through insulation is also a form of heat exchange, although a wasteful one, in that energy with some potential to do work for us is degraded to the ambient level uselessly. Consider the data in Table 8-1,[2] which shows the heat loss from an 8-in. pipe containing a 500°F stream with various thicknesses of insulation starting from zero. Assuming an ambient temperature of 80°F, the available energy loss represented by the heat loss can be calculated from

$$A = Q(1 - T_0/T). \tag{8-6}$$

Pick 2 in. of insulation, for example. From Table 8-1, $Q/A = 92.8$ Btu/ft². An 8-in. pipe with 2 in. of insulation has an outside diameter of about 12.5 in. and therefore has an area of about π ft²/ft length. Therefore,

$Q = 3.15 \times 92.8 = 292$ Btu/ft length,

$A = (292)(1 - 540/960) = (292)(0.43) = 127.5$ Btu/ft length.

To verify this calculation, let us assume the pipe carries steam saturated at 500°F. Over each foot of pipe some steam must condense to supply the heat

Table 8-1

Heat Loss from an 8-in. Diameter Horizontal Pipe with Calcium
Silicate Insulation Containing 500°F Steam[a]

Thickness of insulation (in.)	Surface temperature of insulation (°F)	Heat loss rate per unit area (Btu/h·ft²)	Heat loss rate per unit area compared with that from a bare pipe (%)
0	500	1,770	100
$\frac{1}{2}$	210	321	18.1
1	160	183	10.3
$1\frac{1}{2}$	136	124	7.0
2	122	92.8	5.2
$2\frac{1}{2}$	113	73.2	4.1
3	107	59.9	3.4
$3\frac{1}{2}$	102	50.3	2.8
4	99	43.2	2.4
$4\frac{1}{2}$	95	37.6	2.1
5	92	33.2	1.9
$5\frac{1}{2}$	90	29.6	1.7
6	89	26.7	1.5

[a] From Reference 2. Excerpted by special permission from *Chemical Engineering*
(May 3, 1982). © 1982 McGraw-Hill, Inc., New York, NY 10020.

lost through the insulation. From the steam tables,[1] steam saturated at
500°F has the following properties:

Property	Vapor	Liquid
H (Btu/lb)	1201.7	487.8
S (Btu/lb°R)	1.4325	0.6887

The available energy change due to condensation can be calculated from

$$\Delta A = (H_v - H_l) - T_0(S_v - S_l)$$
$$= (1201.7 - 497.8) - (540)(1.4325 - 0.6887) = 312 \text{ Btu/lb}.$$

The heat lost (ΔH) per foot of pipe is 292 Btu; therefore, the amount of
steam condensed per foot is

$$M = (292)/(1201.7 - 487.8) = 0.408 \text{ lb/ft}.$$

The available energy change (lost work) per foot of pipe is simply

$$(0.408)(312) = 127.5 \text{ Btu/ft}.$$

By either route the result is the same as expected.

Considering economic factors, Table 8.1 clearly shows a point of diminishing returns for insulation. Beyond $3\frac{1}{2}$ or 4 in. of insulation, little percentage improvement in heat loss is obtained. At 4 in. of insulation the surface area of the pipe increases to

$$(16.5/12)\pi = 4.32 \text{ ft}^2/\text{ft length},$$

and

$$Q = (43.2)(4.32) = 186.6 \text{ Btu/ft length},$$

$$- W_{\text{lost}} = \Delta A = (187)(0.43) = 80.4 \text{ Btu/ft length}.$$

To select the optimum thickness when a new plant is being designed, the *capital* and *operating* cost of supplying the lost work must be balanced against the capital cost (insulation) of conserving it. As we shall discuss later, insulation thickness is often chosen on the basis of operating cost alone, assuming the capacity to raise steam or heat process fluids is already available and *paid for*. This may be true for additions to facilities that require no new utility investment, but is certainly not so for grass roots designs. In the general case, then, the optimum insulation thickness should be selected when the desired return on the *net* investment (the cost of insulation minus the cost of the available energy supply) is reached.

II. Expansion–Pressure Letdown ΔP

A. Flow Control and Metering

Flow control and pressure control are most commonly done by taking pressure drop across a valve. Similarly, flow measurement is mainly accomplished by causing a pressure drop across an orifice, venturi, or other restriction. In some meters pressure recovery occurs downstream of the meter, so that the measured pressure difference across the meter is more than the loss measured when normal flow patterns are reestablished. A pressure drop through pipe and equipment has the same net impact on process work requirements.

These operations have several things in common:

(1) the pressure difference has a direct relation to the process lost work;
(2) most of them are isenthalpic, i.e., constant enthalpy operations;

(3) work can be derived from the operation if process and hardware permit;

(4) vapor flows are more significant energy consumers than liquids.

Valves, meters, and other deliberate flow restrictions are generally inserted into the flow stream to accomplish a given purpose. The work lost across the device can generally be calculated in a straightforward manner:

$$-\Delta A = W_{\text{lost}} = T_0 \, \Delta S. \qquad (8\text{-}7)$$

Knowing that these are isenthalpic processes, i.e., the enthalpy of the stream before and after the restriction is the same, allows easy calculation of all properties (generally S and T) after the expansion. Available energy is lost even though enthalpy is conserved. Working a few examples results in some instructive insights.

We have already discussed steam systems at length and shall build on that base to make several points. Take our "standard" high-pressure superheated steam conditions, 1600 psia and 950°F. If the steam is let down across a valve to 600 psia, the enthalpy will remain constant at 1456 Btu/lb, but the entropy will change from 1.5702 to 1.6718 Btu/lb°R[4], and its temperature will be about 890°F. *We can consider this expansion as one through a turbine with zero efficiency.* The work lost through the valve is simply

$$W_{\text{lost}} = (537)(1.6718 - 1.5702) = 54.6 \text{ Btu/lb}.$$

This means that if we could expand the 600-psia, 890°F steam through a perfectly efficient (isentropic) turbine to 77°F ($\sim$0.45-psia) saturated steam, we would get 55 fewer Btu/lb in work produced than if we expanded the 1600-psia steam to the same conditions. This is verified on a Molliere diagram by reading the enthalpies at points on the 0.45-psia line vertically below the starting points for each expansion. This is done as follows:

Property	1600-psia steam	600-psia steam
Initial enthalpy (Btu/lb)	1456	1456
Enthalpy at 0.45 psia (Btu/lb)	840	895
Available work (Btu/lb)	616	561
Change in available work (Btu/lb)	55	

Note that the difference is 55 Btu/lb *only* at T_0. At other temperatures (and pressures) the potential work lost when steam is expanded over a valve is a much greater percentage of the available energy in the 1600-psia steam.

The data are shown in the accompanying expanded enthalpy tabulation (all values in Btu/lb), which was calculated in the same manner as the 77°F table.

Property	1600-psia steam	600-psia steam	ΔW
Initial enthalpy	1456	1456	
Enthalpy at 200 psia	1220	1317	
Potential work	236	139	97
Enthalpy at 100 psia	1162	1244	
Total potential work	294	212	82
Enthalpy at 1 atm	1025	1094	
Total potential work	431	362	69
Enthalpy at 0.45 psia	840	895	
Total potential work	616	561	55

These differences occur at pressures outside the saturation envelope of the Molliere chart, because the turbine exhaust steam is more superheated in the case of the 600-psia inlet. Inside the saturation area, less liquid is formed at the turbine exhaust. In both cases the exhaust steam has higher enthalpy because no work was extracted when the pressure was reduced from 1600 to 600 psia.

B. Pressure Drop

The entropy change associated with pressure changes in a process stream can be calculated by a number of procedures described in textbooks on thermodynamics.[5] When an ideal gas changes pressure, the entropy change is

$$\Delta S_{1 \to 2} = nR \ln(P_1/P_2), \tag{8-8}$$

where n is the number of moles of gas. Thus, ΔS is positive if P_2 is lower than P_1, the initial pressure. If the gas is nonideal, the equation must be corrected by the use of activity or fugacity coefficients. The lost available energy (work, availability, energy) is calculated as before for adiabatic processes:

$$-\Delta A = W_{\text{lost}} = T_0 \Delta S. \tag{8-9}$$

Obviously, the change in entropy can also be evaluated by a thermodynamic equation of state or from tables of data, such as data from the steam tables.

With any of these approaches, a pressure drop in a stream can be related directly to ΔS and ΔA or W_{lost}. Pressure drops are fundamental to many pervasive unit operations. Aside from the obvious applications to fluid flow through pipes, there are pressure drops associated with most processing equipment and metering and control devices. For example, the pressure drop on fractionation trays forms part of the overall change in entropy for a distillation tower, along with the entropy changes associated with heat exchange and the separation being accomplished (see Chapter 7). Similarly, pressure drops associated with control valves and meters result in increases in the entropy of the process fluid and the entire system.

If we consider an isothermal pressure drop in an ideal gas, the equations are directly applicable. For example, if nitrogen drops in pressure from 50 psia to 20 psia, the entropy change for one mole at 537°R is

$$\Delta S = R \ln(50/20) = 1.987 \ln(50/20) = 1.82 \text{ Btu/mol}°R.$$

Under isothermal conditions $\Delta H = 0$, so the change in available energy (work) is

$$W_{lost} = (537)(\Delta S) = 977.3 \text{ Btu/lb} \cdot \text{mol}.$$

If the flow rate is X lb·mol/h, the rate of available energy loss is $977.3X$ Btu/h.

C. Impact of Superheat

If one's purpose is to generate power by passing high-pressure vapor through a turbine, superheating the vapor is very valuable. Consider the following data for 625-psig steam being let down through a 75% efficiency turbine to 135 psig.[1]

1. Inlet Conditions

Temperature (°F)	700	850
H (Btu/lb)	1349	1434
S (Btu/lb°R)	1.579	1.648

2. Isentropic Expansion to 135 psig

H_0 (Btu/lb)	1202	1273
ΔH (work)	147	171

3. Expansion at 75% Efficiency to 135 psig

ΔH (work)	110	128
S_0 (Btu/lb°R)	1.622	1.692
ΔS (Btu/lb°R)	0.043	0.044

Assuming saturated condensate at 135 psig is returned to the boiler ($H = 330.5$ Btu/lb), note that about 8% more fuel [(1434 − 1349)/(1434)] is required to produce 850°F steam as opposed to 700°F steam. However, 16% more work is obtained (128 vs. 110 Btu/lb) as a result of this input, and the remainder is available as enthalpy in the 135-psig steam.

From the second law viewpoint, the lost work in the boiler is reduced because the steam is produced at a higher temperature. More importantly, note that entropy change across the turbine is essentially the same, i.e., the lost work in the turbine is equal, but the amount of work produced is significantly greater. The ratio of work produced to work lost is

$$W_{prod}/W_{lost} = (110)/(537)(0.043) = 4.8$$

for the 700°F case and

$$W_{prod}/W_{lost} = (128)/(537)(0.044) = 5.42$$

for the 850°F case. This represents a significant improvement for a small investment in metallurgy between the boiler and the turbine.

These figures also underline the value of reheating in complex turbine cycles. To improve efficiency steam is often extracted from an intermediate stage of a turbine and resuperheated in the boiler. The two preceding cases represent the benefits from this approach as well. If higher-pressure steam (say 1500 psig) were expanded to a point at which the pressure was 625 psig and temperature 700°F, it could either be allowed to continue through the turbine to exhaust at 135 psig, or it could be extracted and reheated to 850°F and then reintroduced into the machine at a slightly lower pressure. As shown, the last stages of expansion would then produce more work. Of course, these benefits would be partly counteracted by the pressure drop in the reheating system, and some net increase in capital may be required.

III. Mixing

A. Relation to Partial Pressure Changes

The entropy change for a pressure change can be related to that for mixing, as indicated earlier. For ideal gases or solutions, the mixed composition corresponds to a decrease in the pressure of each component from the total pressure of the system to the partial pressure of the component in the mixture. Thus, if pure N_2 and pure O_2 at 1 atm are mixed to form air at 1 atm (79% N_2 and 21% O_2), then the entropy change for the mixing can be calculated from

$$\Delta S = \sum \Delta S_i = \sum n_i R \ln (P/P_i), \qquad (8\text{-}10)$$

where P_i is the partial pressure of each component in the mixture. The same formulation using mole fractions is appropriate for ideal solutions.

For the case in question, the total entropy change per mole of air is

$$\Delta S = (1.987)[0.79 \ln(1/0.79) + 0.21 \ln(1/0.21)]$$

$$= 1.02 \text{ Btu/mol}^\circ R,$$

and $\Delta A = T_0 \Delta S$ if the temperature is held constant. Similarly, when one mole of air is separated into N_2 and O_2, the minimum availability required for the separation is the negative of the work of mixing.

Again, for nonideal gases or solutions, the preceding equation must be corrected by multiplying with activity or fugacity coefficients. In nonideal solutions, therefore, for one mole of solution the entropy change is

$$\Delta S = R \sum X_i \ln \gamma_i X_i, \qquad (8\text{-}11)$$

where γ_i is the activity coefficient for component i and X_i the mole fraction. For gases the composition unit can be changed to partial pressures as before.

B. Different Temperatures

Mixing streams of the same composition but different temperatures also involves an entropy change and its attendant loss of available energy. Consider 1 mol of water at 1 atm and 126°F produced from 0.5 mol at 200°F and 0.5 mol at 32°F. When a material is heated or cooled at constant pressure, its entropy change can be calculated from

$$\Delta S_{1 \rightarrow 2} = nC_p \ln(T_2/T_1).$$

For mixtures this corresponds to

$$\Delta S = \sum n_i C_{p_i} \ln(T_2/T_1)_i. \qquad (8\text{-}12)$$

On cooling ΔS is negative, and on heating it is positive.

To calculate the available energy change for the mixing, both enthalpy and entropy changes must be accounted for. For T_0 equal to 298 K (537°R), this can be calculated from the *changes* in available energy relative to the dead state for both feed streams and the product. Thus,

$$\Delta A_{\text{mix}} = A_{\text{mix}} - (A_{\text{hot}} + A_{\text{cold}}). \tag{8-13}$$

All values are calculated relative to the dead state of $T_0 = 537°R$ by evaluating the difference in the available energy of each stream relative to that of liquid water at $537°R$ ($77°F$):

$$A_{\text{mix}} = \Delta H_{586 \to 537} - T_0 \Delta S_{586 \to 537},$$

$$A_{\text{hot}} = \Delta H_{680 \to 537} - 537 \Delta S_{680 \to 537},$$

$$A_{\text{cold}} = \Delta H_{460 \to 537} - 537 \Delta S_{460 \to 537}.$$

The enthalpy changes are simply calculated from

$$\Delta H = nC_p \Delta T. \tag{8-14}$$

The available energy changes are calculated from

$$\Delta A = nC_p(T - T_0) - T_0 nC_p \ln(T/T_0). \tag{8-15}$$

Therefore, the quantities involved in the mixing of hot and cold water are

$$A_{\text{mix}} = nC_p(586 - 537) - 537 nC_p \ln(586/537),$$

$$A_{\text{hot}} = (0.5)[C_p(680 - 537) - 537 C_p \ln(680/537)],$$

$$A_{\text{cold}} = (0.5)[C_p(460 - 537) - 537 C_p \ln(460/537)].$$

For 1 lb·mol of water, $C_p = 18$ Btu/lb·mol°F.
The available energy of the mixed stream relative to the dead state is

$$A_{\text{mix}} = (1)(18)(586 - 537) - (537)(18) \ln(586/537) = 38 \text{ Btu/mol.}$$

The available energy of the hot stream relative to the dead state is

$$A_{\text{hot}} = (0.5)(18)[(680 - 537) - 537 \ln(680/537)] = 146.4 \text{ Btu/mol.}$$

Similarly, for the cold stream

$$A_{\text{cold}} = (0.5)(18)[(492 - 537) - 537 \ln(492/537)] = 18 \text{ Btu/mol.}$$

Therefore,

$$\Delta A_{\text{mix}} = 38 - (146.4 + 18) = -126.4 \text{ Btu/mol.}$$

All of these calculations could have been made using the steam tables with identical results. In this approach the enthalpy and entropy values for saturated liquid water at the temperatures in question are used. The slight differences in the properties introduced by the fact that the pressures of the liquids at these temperatures are different can be ignored for practical purposes because the pressures are low. From the steam tables:[1]

T (°F)	H (Btu/lb)	S (Btu/lb°R)
32	0.0	0.0
77	45.02	0.0876
126	93.91	0.1748
220	188.13	0.3239

The calculations are made from the definition of available energy, ignoring kinetic and potential energy effects:

$$A = \Delta H - T_0 \Delta S.$$

For the streams in question:

$$\tfrac{1}{18}A_{mix} = (93.91 - 45.02) - (537)(0.1748 - 0.0876), \qquad A_{mix} = 37.14 \text{ Btu};$$

$$\tfrac{1}{9}A_{hot} = (188.13 - 45.02) - (537)(0.3239 - 0.0876), \qquad A_{hot} = 146 \text{ Btu};$$

$$\tfrac{1}{9}A_{cold} = (0 - 45.02) - (537)(0 - 0.0876), \qquad A_{cold} = 18.18 \text{ Btu}.$$

Figure 8-3 shows the available energy losses for the mixing process.[6] Almost all of the available energy of the 220°F stream is lost by mixing it with the 32°F stream. Producing 126°F water by this process is obviously not an efficient step. Thermodynamically, a heat pump would be a far better approach.

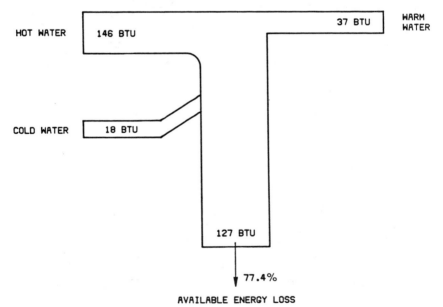

Fig. 8-3   Available energy losses in mixing streams at different temperatures.

IV. Distillation—A Combination of Simple Processes

A. Available Energy Loss

Many studies of the thermodynamic efficiency of distillation have been made, some of which are very sophisticated. Others have addressed the question of separation processes in general, seeking even more fundamental understanding. To this author distillation and other complex unit operations represent combinations of the simpler operations just discussed: pressure drop–rise, heat exchange, and mixing–unmixing. Breaking down the overall unit operation into its components helps in the understanding of where available energy is being lost.

Figure 8-4 shows a typical distillation tower and identifies the suboperations involving the entropy changes that compose it. One tray is used as an example but, obviously, changes occur on every tray. Also, more complex distillation arrangements are possible, such as multiple feeds, interreboilers and condensers, vapor recompressors, partial condensers, etc., but the same component subprocesses still take place in each complication.

As in the previous chapter, the overall available energy loss can be calculated from the differences between the product and feed properties. The losses for major segments of the process were developed in Chapter 7. Further breakdown is possible until each individual cause of entropy production is identified.

Fitzmorris and Mah[7] developed generalized equations for the case of conventional adiabatic distillation, in which the only heat exchange with the environment is through the condenser and reboiler. Further, they excluded heat transfer irreversibilities by defining those parts of the process boundary to be at T_D the temperature of the condenser, and T_B, the temperature of the reboiler, and ignored irreversibilities due to pressure drops. Fundamentally, the minimum (or ideal) work for adiabatic distillation can be formulated as follows:

$$W_{min} = -\sum (mA)_i = (mA)_{dist} + (mA)_{bot} + (mA)_{feed}, \qquad (8\text{-}16)$$

where m_i is the mass flow rate of a given stream and A_i its available energy. The work required to transfer heat Q_D from the condenser at T_D to the environment at T_0 and to transfer heat Q_B from the environment at T_0 to the reboiler at T_B must be added to the minimum available energy change to calculate the total lost work in the tower:

$$T_0 \Delta S = \sum (mA)_i + (1 - T_0/T_B)Q_B - (1 - T_0/T_D)Q_D. \qquad (8\text{-}17)$$

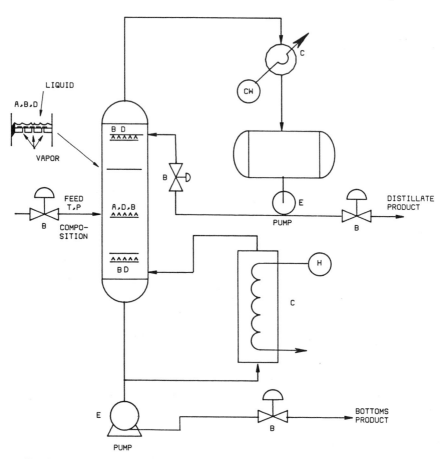

Fig. 8-4 Locating available energy losses in a typical distillation process. A, ΔS for composition mixing–separation; B, ΔS for pressure drop; C, ΔS for heat exchange; D, ΔS for mixing temperature difference; E, ΔS for machinery inefficiency.

Substituting the minimum work term, the relation becomes:

$$T_0 \, \Delta S = (1 - T_0/T_B)Q_B - (1 - T_0/T_D)Q_D - W_{\min}.$$

Thus $(T_0 \, \Delta S)$ accounts for all the irreversibilities that occur because of the mixing of streams at nonequilibrium temperatures and compositions. It can be calculated for each stage of the distillation. Please note again that irreversibilities due to pressure drops have been ignored.

The significance of Eq. (8-18) is that the availability losses at all stages are calculated from the difference between the net work required to transport heat into and out of the tower and the minimum work of the separation.

Thus, any design improvement that reduces this net work difference will reduce the available energy losses of the separation. In practice, this means reducing the heat transport work requirements.

B. Refrigeration Systems—A Direct Relation Between Heat and Work

Closed-cycle propane (or propylene) and ethylene refrigeration systems are used in olefin, butyl rubber, and other plants as low-temperature heat sinks for process chilling and condensing. Methane and multicomponent refrigeration systems (MCR) are used at even lower temperatures in liquified natural gas (LNG) plants, and water chillers are used in air conditioning and for general process applications. At the lower temperatures (below $-40°F$), cascade systems are often used, i.e., heat is rejected from the process to the ethylene refrigerant, which is then condensed by the C_3 refrigeration system, which then rejects the heat to the environment. Thus, the work required to remove a parcel of heat in the ethylene refrigerant range consists of the sum of the work for both C_2 and C_3 refrigerators. A simplified three-stage (C_1, C_2, and C_3) system is depicted in Fig. 8-5, which gives some typical temperature ranges at which services are changed in practice.[8]

If the refrigeration systems could be operated reversibly, the efficiency would be that of a Carnot engine operating between T_{low}, the temperature of the process heat source, and T_0. We have

$$W_{min} = Q_{low}[T_0/T_{low} - 1], \qquad (8-19)$$

where Q_{low} is the heat absorbed at T_{low} and W_{min} the work supplied to a reversible cycle.

Studies have shown that the actual work of real refrigeration systems is directly proportional to W_{min}. As a result, the actual work for a refrigeration system can be calculated by adding an empirical efficiency term to Eq. (8-20):

$$W = W_{min}/\eta = (Q_{low}/\eta)(T_0/T_{low} - 1). \qquad (8-20)$$

The efficiency represents the thermodynamic efficiency of the equipment (machinery, valves, exchangers, selection of refrigerant levels, etc.) in any given system. Improvement of individual components will improve the efficiency of the system.

For cascade refrigeration systems, T_0 is really a correlating variable, not a physical reality. In general, this correction accounts for the temperature differences encountered in the cascade heat exchange. In modern systems the ethylene system heat load is condensed twice and the methane system

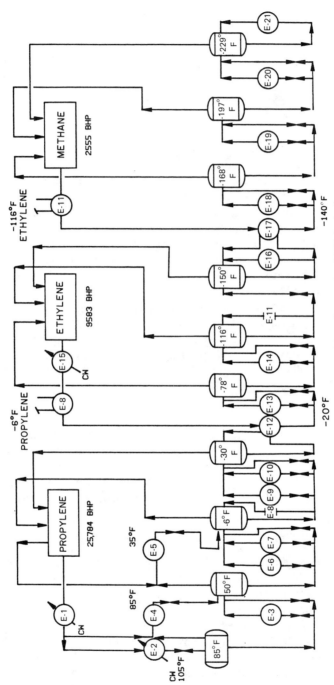

Fig. 8-5 A simple three-stage cascade refrigeration system. BHP, brake horsepower.

three times. This is represented by the higher values of T_0 needed to correlate the actual work.

In a study of a modern refrigeration system, the following values of η were found to correlate well with design and operation data:

Refrigerant	η	T_0 (°F)
Propylene	0.605	105
Ethylene	0.581	125
Methane	0.560	145

The methane system efficiencies were calculated from very limited data and are less reliable. In addition, the figures represent experience for a very specific system and not a general case. For example, if the ethylene condensing pressure is much different from 350 psia, corrections to T_0 of up to 5°R may be necessary.

The significance of these relationships is that the lost work for the system can be approximated for each load on the system. If we define T_{proc} as the mean process temperature in a refrigerated exchanger and T_{ref} as the mean refrigerant temperature, the lost work of the exchanger (reflected directly in compressor shaft horsepower) can be estimated from

$$W_{lost} = (QT_0/\eta)[(T_{proc} - T_{ref})/(T_{proc}T_{ref})]. \qquad (8\text{-}21)$$

This equation assumes that the increment of duty is small enough not to affect the overall efficiency η of the system.

As shown, the work lost might be reduced by changing individual temperatures or by shifting duty to exchangers with more efficient temperature ranges. Note also that for a given value of $T_{proc} - T_{ref}$, the lost work is smaller at higher values of T_{proc} and T_{ref}, so it pays to incur the larger inefficiencies at the highest temperatures in the system.

Given this tool, one can optimize refrigeration systems by building an economic model on top of the thermodynamic one. The economic optimum temperature difference $T_{proc} - T_{ref}$ can be calculated for each exchanger if the capital cost of exchangers and compressors are known along with the cost of power at the plant in question.

V. Combustion Air Preheating

Combustion air preheating provides a classic opportunity to upgrade lower-valued heat to the equivalent of fuel. This is done by providing heat to raise the combustion air temperature from the ambient temperature to higher values using waste heat rather than the heat of combustion as is the usual situation. Consider the chemical equation (7.14) in Chapter 7 for the combustion of methane: $11\frac{1}{2}$ mol flue gas are produced by the combustion of 1 mol CH_4 and 10.5 mol air. The heat liberated is 345,163 Btu/mol CH_4.

If the air and fuel enter the process at 77°F, the temperature of the flue gas at the completion of the reaction but before any cooling of the flue gas occurs is the adiabatic flame temperature. It can be approximated by a heat balance assuming a constant heat capacity of 7.5 Btu/mol:

$$Q_R = (11.5)(7.5)(T_f - 77) = 345,163 \text{ Btu/mol},$$

$$T_f = (345,163)/(11.5)(7.5) + 77 = 4002 + 77 = 4079°F.$$

If the combustion air were preheated to 477°F by an external heat source, the heat available to the flue gas would be increased by the amount of the preheat:

$$Q_{ph} = 10.5 \times 7.5 \times 400 = 31,500 \text{ Btu/mol}.$$

The amount of available energy expended to heat the air will depend on the source of heat. If it were 900-psia steam (which has a saturation temperature of 532°F), 47 lb would be needed to provide 31,500 Btu. The available energy contained in each pound of steam can be calculated from data given in the steam tables to be 444 Btu/lb. The available energy used to raise the air temperature would be $47 \times 444 = 20,868$ Btu/mol CH_4. If the air were heated by cooling flue gas from 532°F to 132°F, one might evaluate the available energy at the mean temperature of the flue gas, 332°F (792°R), as follows:

$$A = (31,500)(1 - T_0/T) = (31,500)(1 - 537/792)$$

$$= 10,142 \text{ Btu/mol } CH_4.$$

A lower expenditure of available energy would be expected in this case because not all the heat is transferred at the highest temperature, as it was in the saturated steam example. Both of these approaches will have the same benefit (as calculated below), so that the lower available energy input route is preferable.

Using preheated air will raise the flame temperature. This is calculated by adding Q_R and Q_{ph} and repeating the given procedure:

$$Q_{ph} + Q_R = 31,500 + 345,163 = 376,663 \text{ Btu/mol},$$

$$T_f = (376,663)/(11.5)(7.5) + 77 = 4444°F.$$

The available energy of the flue gas is increased because both the flame temperature and the heat content are higher. By the Carnot ratio at the mean temperature between T_f and T_0 for the case of no preheating,

$$A = (345,163) \left(1 - \frac{T_0}{T}\right) = (345,163) \left(1 - \frac{(537)(2)}{(4539 + 537)}\right)$$

$$= 272,132 \text{ Btu/mol CH}_4.$$

With preheating,

$$A = (376,663) \left(1 - \frac{(537)(2)}{(4904 + 537)}\right) = 302,313 \text{ Btu.}$$

The available energy of the flue gas is increased by 29,114 Btu/mol CH_4, or almost the entire quantity of the air preheat. There is a significant upgrading of the available energy impact of this quantity of heat. In practice, this is usually taken advantage of by reducing the amount of fuel burned.

Summary

In this chapter the fundamental unit operations of heat transfer, pressure drops and rises, and mixing are discussed in detail. More complex processes such as refrigeration and distillation can be pictured as combinations of these simpler operations. The opportunity to upgrade the available energy contribution of a parcel of energy is demonstrated by an example of combustion air preheating. In this approach the vast majority of the heat supplied to the combustion air is recovered as available energy in the flue gas, even though the available energy input is only a fraction of this amount, depending on the source.

Notes to Chapter 8

1. J. H. Keenan, and F. G. Keyes, "Thermodynamic Properties of Steam." Wiley, New York, 1936.

2. M. McChesney and P. McChesney, Insulation without economics. *Chemical Engineering* **89,** 77 (1982).

3. "Relevance of The Second law of Thermodynamics to Energy Conservation," DOE/CS/ 40178-000-01. A report prepared by the National Bureau of Standards and General Energy Associates, Inc. of Cherry Hill, New Jersey, January, 1980. U. S. Govt. Printing Office, Washington, D.C.

4. From Reference 1:

Conditions	H (Btu/lb)	S (Btu/lb°R)
600 psia, 900°F	1462.5	1.6762
600 psia, 800°F	1435.2	1.6558
Difference	27.3	0.0204

To interpolate, $1462.2 - 1456.6 = 5.9$ and $5.9/27.3 = 0.218$; therefore,

$$S_{600\,\text{psia}} = 1.6762 - (0.218)(1.6762 - 1.6558) = 1.6718.$$

5. B. F. Dodge or M. Sussman, for example. B. F. Dodge, "Chemical Engineering Thermodynamics." McGraw-Hill, New York, 1974. See also Reference 9.

6. G. E. Yeo, Thermodynamic Analysis of Refrigeration Systems. Personal communication, August, 1981.

7. R. E. Fitzmorris and R. S. H. Mah, Improving distillation column design using thermodynamic availability analysis. *American Institute of Chemical Engineers Journal* **26** (2), 265 (1980).

8. A. J. McCarthy and M. E. Hopkins, Simplify refrigeration estimating. *Hydrocarbon Processing* **50,** 106 (1971).

9. Adapted from M. V. Sussmann, "Availability (Exergy) Analysis." Mulliken House, Lexington, Massachusetts, 1980.

9

Use of Thermodynamic Analysis to Improve Energy Efficiency

The essence of knowledge is, having it, to use it.
— Confucius

Introduction

Having carried out the considerable arithmetic described in the previous chapters, the challenge still remains to turn the information developed into a profit. Ideally, the new insights and understanding resulting from thermodynamic analysis can be used to reduce energy consumption in new and existing facilities through design updates, and they may even save some capital cost for the facilities involved. Improvements might be expected in the following areas:

(1) improved process and equipment designs for onsite units and energy systems,
(2) avoidance of design mistakes and unsound technical proposals,
(3) more intelligent process selection for new units,
(4) more energy efficient process developments,
(5) improved research guidance.

In addition to these, a number of people are trying to develop quantitative systems that would lead the process designer to more optimal process and equipment designs in a very systematic way. These calculational systems will not be discussed in this chapter, but rather will be explored later in the book because of their complexity.

The relationships between energy efficiency and capital cost will be discussed in Chapter 10. Intuitively, we may feel that if much less fuel need be consumed and less energy transported to the process units, then surely less hardware will be required in the plant to accomplish the necessary energy inputs. This should result in a lower overall capital cost to the plant in question. However, energy conservation programs to date have generally involved increased investment rather than less. Some global energy policy studies and the sophisticated design systems mentioned previously tend to support the intuitive conclusion.

This chapter will discuss a systematic approach to using the data developed in thermodynamic analysis to improve the energy efficiency of existing and new facilities. As discussed earlier, there are no guarantees that simply carrying out a thermodynamic analysis will provide the creative insight needed to improve a process or equipment design. Sound engineering analysis and creative thinking are still required. The approach used is to discuss a number of examples in which improvements have been suggested as a result of second law analysis. The examples embody a variety of approaches and processes, but the net result will not be a cookbook for success. Hopefully, the reader will develop a feel for the strategy of turning a particular insight into profits.

I. Overall Strategy

Thermodynamic analysis provides two separate avenues for the pursuit of energy efficiency improvements. These are

(1)   reducing the ideal work for the desired change,
(2)   reducing the lost work of the process selected to accomplish this change.

The approaches have different characteristics and limitations.

Changing the ideal work of a process can involve only a limited number of variables. Theoretically, the ideal work is independent of the process chosen and depends only on the initial and final states of the raw materials and products. Thus, the variables to be manipulated are confined to the initial and final conditions and the initial and final compositions. In some cases changes in the mass flow rates are more related to ideal work changes than they are to lost work changes, and might be considered to belong in the ideal work category.

By their nature reductions in ideal work requirements are generally strategic in nature, result from an overall view of the process, and require less data than reductions in lost work. Also, they are by nature more difficult to find and generally require more radical departures from existing technology, resulting in a difficult "sell" to company management.

These difficulties should not inhibit the search for improvements. Relatively little manpower is required to carry out the analysis, and even small

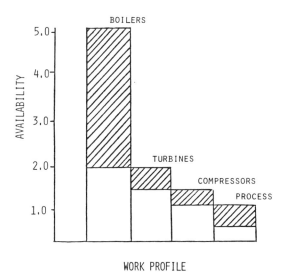

WORK PROFILE

Fig. 9-1   Minimum (□) and lost (▨) availability in the purification section of an ethylene plant: lost availability breeds lost availability.

changes generally produce significant results. Remember that many commercial processes have overall thermodynamic efficiencies in the range of 10 to 20%. As a result, small improvements in ideal work requirements have large multipliers back through the energy system.

Figure 9-1 shows the available work relationships in the purification section of an ethylene plant. Small improvements in each stage of the process cascade back to the boilers, where a multiplier of about five on fuel consumption is observed. This is true because of the energy system cascade.

If the lost work of the separation process is reduced in such a way as to reduce refrigeration requirements, in effect the ideal work of the refrigeration section has been reduced. This is true because the only product of the refrigeration section is to supply the available energy needed to cool the process in question. The ideal work of the refrigerant compressor is reduced, and with it the lost work encountered in that step. Working backward through the power supply train, the actual work for the compressor becomes the ideal work for the turbine that drives the compressor. Similarly, the actual work for the turbine is the ideal work for the boiler and steam system. All along the chain, any change in ideal work results in a change in lost work in the preceding process segment. The magnitude of each lost work change will depend on the efficiency of the particular step. Compressors and drivers have high thermodynamic efficiencies, but boilers do not. Thus, a small change in refrigeration load is magnified substantially at the boilers, as depicted. The title of the graph says it well, "lost availability breeds lost availability."

Please note that the real source of the improvement was a reduction in the work lost in a process segment. This is an indication of where most improvements will be found.

Initially, the level of detail required for analyzing the ideal work is relatively simple. Only when some opportunities are uncovered should more detail be considered. As indicated earlier, one needs to keep an open mind about the possibilities for improvement.

A. Stepwise Approach

Thermodynamic analysis can develop a plethora of data that can overwhelm the engineer. The general approach is to start simply and add detail as avenues of inquiry open up. A stepwise approach is shown in the following list.

Stepwise Approach

Perform the thermodynamic analysis
Generate ideas
Prioritize ideas
Assess the technical feasibility
Create a new material balance
Rework the thermodynamic analysis at the appropriate level of detail
Evaluate the benefits versus the cost of the development program

---

Obviously, the first step is to carry out a thermodynamic analysis. Two factors are important in this analysis. The first is the amount of lost work associated with each process step, and the second is the overall level of efficiency of the particular process step. Steps that include large lost work terms but operate at high efficiencies generally are not subject to much improvement with conventional technology. Large centrifugal compressors are good examples of this type of process segment.

Having developed a simple table of lost work and efficiency terms, the next step is to generate ideas for process and/or equipment improvement. The same approach outlined in the discussion of using energy audit data is indicated, namely, a brainstorming session and the testing of all assumptions inherent in the initial process design. An uninhibited list of ideas needs to be generated. Often there will be insufficient data to make any analysis of whether a particular idea is likely to be fruitful. In these circumstances a further breakdown of the thermodynamic analysis of the area in question is usually warranted.

When the idea flow has run dry for the moment, a priority list should be drawn up. The priorities are based on the magnitude of the lost work terms and the potential for improvement (as measured by the current second law efficiency of the step being considered).

The technical feasibility of each idea is then tested. This is generally done by developing a practical embodiment of the idea. Thus, if one idea is for a catalyst to reduce reaction temperature from 1500°F to 500°F, one can peruse the literature for any hint that such a catalyst might be available. Or, if one hypothesizes that a topping cycle should be added to the convection section of the boilers, it can be determined whether a suitable working fluid and machinery are commercially available.

The next steps are to rework the material balance, and possibly the thermodynamic analysis, to determine the magnitude of savings that accompanies the practical embodiment of the idea. Here again some additional detail may be required to establish the practical benefits that might be achieved.

The last step, as always, is an assessment of the cost–benefit ratio for the most promising ideas on the list. Entropy plays no part in economic analysis, except to make sure that savings are properly valued. The important factors will be investments required in process development, new facilities, and management time, plus an assessment of the probability for success. These are the same elements involved in any evaluation of any development or investment in new facilities. Conventional engineering approaches are more than adequate.

The goal of this systematic procedure is to aid the engineer in achieving the creative leap from data to profits. It is a framework to guide the engineer's thinking and focus it on areas with the most potential.

B. Considering the "Classic" Example

For years steam power plant engineers have analyzed their process from a second law viewpoint. Perhaps this is because the process embodies the essence of the second law, namely, the transformation of heat to work. Perhaps it is because the data were readily available. In any event, various aspects of steam processes have been subjected to second law scrutiny for some period of time and represent simple examples of the difference be-

Table 9-1

Enthalpy and Available Energy Losses for a Steam
Power Cycle

Parameter	Enthalpy (Btu)	Available energy (Btu)
CH$_4$ burned	345,163	344,502
Electricity produced	83,334	83,334
Total loss	261,829	261,168
Breakdown of losses		
Stack	45,109	Including boiler
Condenser	216,687	9,457
Turbine	0	26,555
Boiler	0	226,446
Pump	0	Neglected
Total loss	261,796	262,458
Overall efficiency	24.1%	24.2%

tween the first and second law viewpoints and of the generation of ideas for improvement.

In Chapters 6 and 7 a simple steam power cycle was analyzed. The results of the lost work analysis are shown again in Table 9-1. In this process the desired product is the electricity produced by the turbine. The objective of the process is to maximize the electric power recovered from the burning of a given amount of fuel. Also shown in Table 9-1 is the first law analysis of the process. In this analysis the enthalpy discarded at each stage of the process represents the loss. The quantity of product electricity is identical in both cases.

The contrast between the two forms of analysis is striking. From the first law standpoint, the major losses are in the condenser and the flue gas from the boiler. From the second law standpoint, the major losses are in the combustion process and in the turbine. Following the guidance of the first law, the energy conservation engineer would look around his plant for low-temperature energy sinks to recover the heat rejected in the furnace stack and condenser. The energy conservation efforts guided by the second law, however, would focus on two main areas: reducing the lost work in the furnace and recovering more work at the turbine. The second law engineer would not ignore opportunities for recovering heat from the flue gas or from the condenser, if such is possible, for they are inherent in achieving the given objectives.

The second law objectives represent a combination of ideal work and lost

Table 9-2

Ideas for Improving Energy Efficiency

Enthalpy approach	Available energy approach

1. Recover heat lost to cooling water in condenser with low-temperature sink offsite ($\sim 90°F$)
 2. Recover stack heat, consider air preheating
	1. Consider cogeneration opportunities for condenser heat and be open to changing turbine discharge pressure
	 2. Reduce amount of fuel fired through air preheating, use stack heat
	 3. Increase amount of electricity generated at constant fuel
	    a. Increase steam pressure
	    b. Study superheat temperature
	    c. Improve turbine efficiency
	    d. Consider bottoming cycle on stack
	 4. Add topping cycle between combustion and steam temperature

work considerations. The ideal work for the process is the maximum available energy that can be derived from the fuel burned. Improving the overall efficiency of the process would move the actual work obtained from the process closer to the ideal. Conversely, reducing the amount of fuel fired to produce the same amount of electricity would be a reduction in ideal work for the process.

Priorities generated by a first law approach involve only the quantity of heat to be recovered. Thus, the condenser is the first priority because of the large quantity of heat lost, even though it is available at low temperature (about 90°F). The only other loss recorded in the analysis is from the furnace stack, where about 25% as much heat is rejected but at a much higher temperature. Very little of the condenser heat can be used onsite. If the boiler feed water makeup is very cold, some energy might be recovered, and in most places a small amount of this heat might be used to preheat combustion air. As a result, the first idea generated and listed in Table 9-2 is to find a low-temperature sink offsite that might use this relatively large quantity of warm water. This might include heating for a commercial building, a greenhouse, or a shrimp farm, or some other low-temperature application. Recovered stack heat obviously has considerably more potential uses, because the energy is at a relatively high temperature. Air preheat is self-contained within the boiler area, which would be the first place to look. About 300°F air could be achieved by cooling the stack gases from 600° to 300°F.

The first two items in the available energy column of Table 9-2 involve the same steps. The difference in viewpoint is reflected in the choice of words, however. The focus is on reducing the amount of fuel fired and maximizing the power generated from the necessary fuel consumption.

Most likely, the series of ideas listed in items 3 and 4 would only have originated from a second law viewpoint. Increasing the steam pressure and the superheating temperature are obvious ways to recover more of the desired product (electricity) from a given amount of fuel. Without exotic technology and without changing the stack temperature or the components of the process in any way, these two ideas use the available energy of the fuel more effectively. Obviously, the capital cost of the steam system will increase, but this will be partially offset by smaller-sized piping and a smaller heat exchanger and cooling tower. The improvements are essentially additive to those suggested in items 1 and 2.

The ideas listed as items 3c and 3d follow directly from the same thinking, with varying degrees of technology "step-out." Improving machinery efficiency and adding a bottoming cycle simultaneously would provide more product power. These items involve changes in machinery and process technology and therefore may not be immediately applicable. Item 4 is a "classic" thermodynamic suggestion: add a topping cycle. However, it should be realized that mercury vapor topping cycles were once in commercial operation in conventional steam power plants about 50 years ago.

At the very least, these last three ideas represent some guidance for development activities in a longer-range program for improved energy efficiency.

Note that the thermodynamic analysis presented for the process is very simple. None of the details for the boiler feed water preheating system were presented, nor were any representations made of the details of the turbine and condenser systems. If more details were provided a more detailed analysis could have been carried out, with more specific suggestions for improvement generated. On a first law basis, however, additional process details would provide little food for thought unless further losses to the atmosphere were identified.

C. Considering the Potential for True Changes in Ideal Work

As indicated earlier, the ideal work can only be changed if the initial and final states of the process materials are changed. This could include pressure, temperature, physical state, and composition. In some cases it might also include the flow rates of any of the materials.

In an ethylene plant the pure product is generally exported to tributary processing units at about 1500 psig, often as a gas. This high pressure is often required for transmission of the product by a major pipeline, but often some part of the product goes to a nearby captive polymerization unit. Conventional low-density polyethylene plants operate at very high pressures (about 50,000 psig), and the higher the pressure at which the ethylene feed arrives, the less compression is required in the polymer plant. With the advent of new technology (Unipol and other), the polymerization reaction takes place at only a few hundred pounds per square inch and is much more comparable to the pressures at which pure ethylene is produced in the final product tower of the ethylene plant. Therefore, in certain cases it might be appropriate to reduce the pressure of the pure ethylene product to save energy of compression.

The savings in ideal work for this step can be estimated rather simply. From Chapter 7 the ideal work for any process can be expressed as follows:

$$W_{min} = \Delta H - T_0 \, \Delta S. \tag{9-1}$$

As typical conditions, we shall consider the ideal work associated with the compression of ethylene vapor at 100°F from 25 to 100 atm. Ethylene is not quite an ideal gas. From Exxon data sources the enthalpy of ethylene decreases 8 Btu/lb, or a little more than 1% on compression from 25 to 100 atm. The entropy change could also be estimated directly from the data, but it may be more instructive to use the ideal gas equation for the entropy change as a function of pressure [Eq. (8-8)], because the small enthalpy change indicates that the nonideality of ethylene is not large. The ideal gas entropy change can be calculated from

$$\Delta S = nR \ln(P_1/P_2). \tag{9-2}$$

The molecular weight of ethylene is 28, so the entropy change for 1 lb of product being compressed from 390 psia to 1515 psia can be expressed as follows:

$$\Delta S = (1.987/28) \ln(390/1515) = -0.0964 \text{ Btu/lb.}$$

The ideal work change can be calculated as follows:

$$W_i = \Delta H - T_0 \, \Delta S = -8 - (537)(-0.964) = 44 \text{ Btu/lb.}$$

This relatively small quantity of work represents about 2% of the ideal work for the entire process. In terms of actual work saving, the efficiency of the compressor and driver are among the highest of any units in the plant, about 65%; therefore, the actual work saved in the machines themselves will not increase by a large factor. If the driver for the compressor is a steam turbine, then a three times larger saving in boiler fuel can be obtained.

A reasonable conclusion from this discussion is that a rather significant change in product pressure has a rather small, but measurable effect on the ideal work of the process.

D. Purity Impact

Ethylene is generally produced at a purity in the range of 99.9%. If product purity were reduced to 99.0%, would this have a significant impact on the ideal work for the process?

For such small changes in concentration, enthalpy effects can be assumed to be zero. The ideal work change is then represented by

$$W_i = -T_0 \, \Delta S. \tag{9-3}$$

Ethylene approaches an ideal gas. Changes in composition can then be considered as changes in the partial pressure of the gas in a mixture. Thus, the entropy change of purification can be calculated in much the same way as the minimum work for separation or the entropy change for a change in absolute pressure of the pure material. Thus,

$$\Delta S = nR \, \ln(P/P_i) = (1.987/28) \, \ln(1/0.99)$$

$$= 0.0007 \text{ Btu/lb}^\circ\text{R},$$

$$W_i = 537 \times 0.0007 = 0.38 \text{ Btu/lb}.$$

As demonstrated in Chapter 7, the thermodynamic efficiency of fractionation processes are very low. If one assumes an efficiency of about 3%, the change in actual work that may occur due to this change in ideal work would be about 12 Btu/lb. This represents a very small percentage of the 1800 Btu/lb needed to produce ethylene from feed by the steam cracking process.

For larger changes in purity the actual work associated with purification can become significant. A practical case often confronted is whether to produce chemical grade propylene (93%) or polymer grade, which is generally 99 + % propylene. A similar analysis demonstrates that this composition change is significant. Propylene has a molecular weight of 42. Therefore,

$$\Delta S = (1.987/42) \, \ln(1/0.93) = 0.0034 \text{ Btu/lb}^\circ\text{F}$$

$$= 537 \times 0.0034 = 1.8 \text{ Btu/lb}.$$

Again, if the separation efficiency is typically low (about 3%), the actual work saved by this small change in ideal work is approximately 60 Btu/lb. This magnitude starts to approach the level encountered for the major product pressure change discussed earlier.

The purpose of the foregoing discussion was to demonstrate that the impact of potential changes in ideal work can be scoped out quickly and simply. If any of these changes look to be significant, a more detailed analysis would have to be made. In the examples cited a complete analysis of the entire purification and steam power system would probably be required to establish the full impact on the actual work requirement for the process.

II. Reducing Available Energy (Work) Losses

A. Process Design – Selection Improvements Using Proven Technology

Let us return to the fractionation problem used in Chapter 7 to demonstrate the methodology of thermodynamic analysis. Figure 9-2 recapitulates

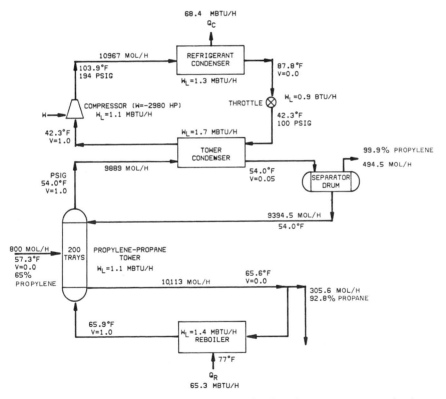

Fig. 9-2 Energy flow diagram for a refrigerated fractionation tower. V, vapor fraction. From Reference 1.

that analysis, showing the lost work terms for each process step. A mixture of propane and propylene is to be separated into relatively pure product. The proposed process would do this in a refrigerated distillation tower using free heat from the environment at 77°F for the reboiler. The only work required for the system is that of the refrigeration compressor that provides coolant for the tower condenser. You will recall that the ideal work for this separation is very low relative to the actual work performed, giving an overall thermodynamic efficiency on the order of 3%. Consider yourself challenged with the problem of suggesting process improvements for this unit in your plant. Sit down for a few moments to make a list of potential ideas for process improvement based on the data provided in Fig. 9-2. Remember the approach described earlier: test all assumptions and match sources and sinks.

Lost work is encountered in every segment of the process, and there is no overwhelming contributor on which to focus attack. Let us spend some time analyzing the assumptions that are inherent in the proposed process. First, is fractionation the proper separation technique? Certainly current membrane separation processes are not well suited for removing either species of the feed stream from the other. One could consider a chemical separation in which the olefins were reacted or polymerized out of the feed stream directly, assuming the product was desired. This amounts to having the downstream processing section take in more dilute feed, and it may or may not be feasible. Basically, we are asking whether the separation needs to be done at all. Beyond these comments we have nothing to contribute in the way of a practical process idea that would replace fractionation as the separation process for this job.

Given a fractionation process, there are still assumptions to be explored. In the proposed process sequence, advantage is taken both of the "free" availability used to drive the reboiler and of the fact that at low tower pressures the relative volatility of the separation is improved. The price paid for these benefits is the need for a refrigerated condenser and a lot of hardware. Can the benefits be preserved but the price reduced? Most losses (5.0 out of 7.5 MBtu/h) occur in the overhead circuit involving the tower condenser, the refrigeration condenser, and the throttle valve. The compressor itself is relatively efficient (about 65%). To maintain the low pressure some compression is required so the heat rejection necessary to provide reflux can be accomplished.

Two approaches can be postulated at the outset: a heat pump and higher tower pressure. In the heat pump scheme the overhead vapors from the tower would be compressed to the point where they condense at a high enough temperature to reboil the tower. The excess heat needed to maintain the enthalpy balance could be rejected at the compressor discharge, where temperatures are greater than 77°F. This possibility is shown in Fig. 9-3. It eliminates the tower condenser, the refrigerant condenser, and the throttle valve, but still maintains the desired low pressure level for the tower itself. You will note that this idea has the added advantage of eliminating considerable hardware, and thus also holds the potential for reducing the capital cost of the separation.

The higher-pressure option would allow operating the condenser with cooling water. This would mean that a higher-temperature heating medium would be needed for the reboiler, and the tower would have to be physically larger because of the debit in relative volatility for the separation that is inherent in operation at higher pressure. However, relatively low-level heat would still be required for the bottom of the tower at these pressures, and the availability consumption might be smaller with this process route as well (see Fig. 9-4).

Simplified descriptions of both process alternatives are shown in the accompanying list.

Possible Fractionation Process Improvements

Install heat pump on tower overhead
Eliminate refrigeration system by increasing tower pressure
Replace throttle valve with expansion engine
Install Carnot engine to recover work from tower temperature difference (66°F – 54°F)
Install Carnot engines on heat exchangers to produce work from the temperature difference between process and coolant

---

Both processes involve conventional technology, and process design should be straightforward. A 20-psig steam reboiler was assumed in the high-pressure case, but lower pressures would also be possible with the bottoms temperature shown.

The other items listed are more "classical thermodynamics" in nature. Replacing throttle valves and installing Carnot (heat) engines is theoretically possible, but not practical at this time. Thus, with the assumption that we cannot substitute for fractionation, the hope for process improvement boils down to installing a heat pump system or raising the tower pressure.

An analysis of the three alternatives is shown in Table 9-3. The available energy lost for each segment of the process is listed. The heat pump case provides a significant improvement in available energy consumption. In addition, appreciably less hardware is required than for the refrigerated tower, and a lower capital cost is also likely.

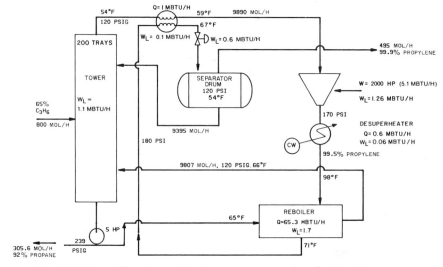

Fig. 9-3   Energy flow diagram for a heat pump system for a fractionation process.

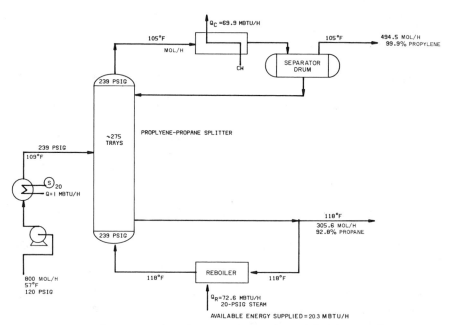

Fig. 9-4   Energy flow diagram for a higher tower pressure system for a fractionation process.

Table 9-3

Thermodynamic Analysis for the Propylene–Propane Separation Process

	Lost work (MBtu/H)		
Component	High-pressure option	Refrigeration option	Heat pump option
Tower	2.0	1.1	1.1
Tower condenser		1.7	—
Tower reboiler	18.3	1.4	1.7
Compressor	—	1.1	1.26
Desuperheater	—	—	0.06
Refrigerant condenser	—	1.3	—
Throttle valve	—	0.9	0.6
Superheater	—	—	0.1
Total lost work	20.3	7.5	4.82
Theoretical work	0.3	0.3	0.3
Total lost work plus theoretical work	20.6	7.8	5.12
Actual work	20.6	7.6	5.1

Some comments on the high-pressure tower should also be made. The available energy consumption for the 20-psig steam does not depend on whether it originates in a boiler or as turbine exhaust. The amount of fuel chargeable to steam depends on its source, but the available energy consumed can only be changed by lowering the temperature of the heat source. Obviously, this can be done. If a 12°F temperature difference were provided for the high-pressure tower (as for the refrigerated case), then 130°F steam could be used. With a latent heat of 1020 Btu/lb, some 71,000 pph would be required. The available energy consumed would be calculated from

$$\Delta A = (71,000)[(H_{130} - H_{T_0}) - T_0(S_{130} - S_{T_0})]$$

$$= (71,000)[(1117.9 - 45) - (537)(1.9112 - 0.088)]$$

$$= 6.64 \text{ MBtu/h}.$$

Therefore, the high-pressure tower can approach the available energy consumption of the low-pressure tower if the optimum heat source is available. This is rarely the case, but it should be explored before a final decision is made.

A point worthy of note is that the entire analysis presented here, including the original lost work analysis (but not the data preparation), took less than one day to do using only a hand calculator.

B. Isopropyl Alcohol Drying Process Selection

The objective of this example is to demonstrate how the energy effects of process selection can be simplified by thermodynamic analysis and related to the utility pricing system discussed earlier in the text. For various product quality reasons, a need existed for improving the process used for the dehydration of isopropyl alcohol. Because isopropyl alcohol and water form an azeotrope, extractive distillation is the conventional route for upgrading 91% isopropyl alcohol to anhydrous material. The two processes under consideration are shown in Fig. 9-5 and 9-6, and the rudimentary material balances for both processes are shown in Table 9-4. The processes contain identical elements. These involve preheating the feed entering an extractive distillation tower, where an entrainer is used to break the azeotrope. The alcohol is recovered as a pure product at the bottom of the tower. The overhead from the tower is a mixture of water and the entrainer, which is then separated in a second tower, the water being rejected and the entrainer recycled. The differences between the two processes involve the entrainer, the tower pressures and temperatures, and the utility supply system. Figure 9-5 shows the balance for a low-temperature entrainer that is a 20-psig steam consumer and uses cooling water on the condenser. Figure 9-6 uses 175-psig steam in the reboiler but recovers 20-psig steam in the overhead from the first tower. In both cases 20-psig steam is used to reboil the entrainer recovery tower in about equal amounts. The low-temperature

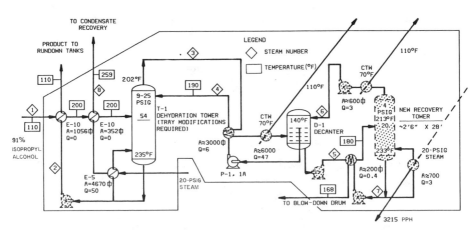

Fig. 9-5 Low-temperature entrainer process used to upgrade 91% isopropyl alcohol to anhydrous material. All heat duties are in million Btu per hour, and the analysis is based on 55 million gal/yr anhydrous isopropyl alcohol. The shaded areas represent new equipment. Φ, heat exchange area in square feet; Q is in MBtu/h. From Reference 1.

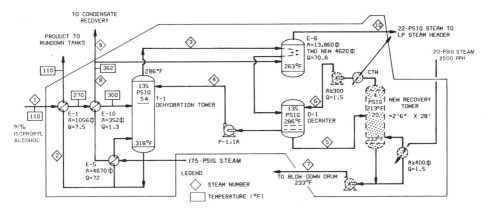

Fig. 9-6 High-temperature entrainer process used to upgrade 91% isopropyl alcohol to anhydrous material. All heat duties are in million Btu per hour, and the analysis is based on 55 million gal/yr anhydrous isopropyl alcohol. The shaded areas represent new equipment. Φ, heat exchange area in square feet; Q is in MBtu/h. From Reference 1.

entrainer process involves the net consumption of 57 pph of 20-psig steam, whereas the high-temperature process consumes about 94 kpph of 175-psig steam but produces 80 pph of 20-psig steam after allowing for the steam required in the recovery tower.

Given a set of steam values for the plant in which the process is to be located, one can quickly determine the economics of each process by assuming all the other items are equal. However, an available energy balance would identify whether one process has an inherent longer-term advantage over the other and indicate whether the plant steam values are correct.

Before doing this, however, it is instructive to calculate the ideal work for the process. Because the objective of the process is the same, the ideal work is the same for both process schemes. In the calculation to follow, it is assumed that isopropyl alcohol and water form ideal solutions. If that is not the case, then fugacity coefficients must be applied to each term in the equation to correct for nonideality. As before, the key element is to determine the entropy change for the process. In separations of this nature, the negative of the entropy change of mixing is a measure of the entropy change of separation. For ideal solutions the enthalpy effects of mixing are zero, and we shall assume that they are negligible for the process in question. As will be shown, the ideal work of separation is on the order 600 kBtu/h for a 50 million gal/yr capacity.

The ideal work of separation is the negative of the work of mixing. We start with

$$W_i = -T_0 \, \Delta S_{mix}. \tag{9-3}$$

Table 9-4

Material Balance for the Low-Temperature (Fig. 9-5) and High-Temperature (Fig. 9-6) Entrainer Processes[a]

Flow rate (pph)	Steam number									
	①	②	③	④	⑤	⑥	⑦	⑧	⑨	⑩
Figure 9-5										
Low-temperature entrainer			119,452	119,452	71	71				
Isopropyl alcohol	43,481	43,481	49,148	49,148	1,890	1,890				
Water	5,992	4	13,813	7,825	8,240	2,252	5,988	53,589		
Total	49,473	43,485	182,413	176,425	10,201	4,213	5,988	53,589		
Figure 9-6										
High-temperature entrainer			177,617	177,617	1,177	1,177				
Isopropyl alcohol	43,481	43,481	120,887	120,887						
Water	5,992	4	29,479	23,492	6,776	810	5,987	93,500	11,660	
Steam										81,840
Total	49,473	43,485	327,983	321,996	7,953	1,987	5,987	93,500	11,660	81,840

[a] All values in pounds per hour. From Reference 1.

Assuming the ideal binary solution, we have

$$\Delta S_{mix} = R\left[y_1 \ln \frac{1}{y} + (1 - y_1) \ln \frac{1}{1 - y_1}\right], \tag{9-4}$$

where y is the mole fraction and R the gas constant, 1.987 Btu/mol°R. For 91% by weight isopropyl alcohol solution, the molar breakdown is

$$91/60 = 1.52 \text{ mol isopropyl alcohol,}$$

$$9/18 = 0.50 \text{ mol } H_2O,$$

$$1.52 + 6.50 = 2.02 \text{ mol solution,}$$

$$y = 0.7525.$$

Considering Fig. 9-6 for 43,481 pph isopropyl alcohol, the molar flow rates are 724.7 mol/h isopropyl alcohol and 332.9 mol/h H_2O, for a total of 1057.6 mol/h. The change in entropy is

$$\Delta S/\text{mol} = (1.986)[0.7525 \ln(1/0.7525) + 0.2475 \ln(1/0.2475)]$$

$$= (1.986)[(0.7525)(0.284) + (0.2475)(1.396)]$$

$$= 1.11 \text{ Btu/lb·mol solution,}$$

$$\Delta S_{tot} = (1.11)(1057.6) = 1174.8 \text{ Btu/h.}$$

The ideal work of separation when $T_0 = 530°R$ is

$$W_i = -(530)(1174.8) = -0.623 \times 10^6 \text{ Btu/h.}$$

C. Calculating Actual Availability Changes

Because the feed and prime product from each of these processes are identical, the thermodynamic analysis can be simplified by ignoring the isopropyl alcohol altogether. This leaves us dealing only with the water and steam streams that cross the boundaries indicated by the solid lines in Figs. 9-5 and 9-6.

The actual work requirements will be calculated from the total change in availability across the process. For the low-temperature entrainer process of Fig. 9-5, we make the following simplifying assumptions:

(1) ignore the effect of the cooling water;
(2) assume the input and output of isopropyl alcohol cancel because the temperatures are equal;
(3) ignore the effect of the motors.

Table 9-5

Availability Data for the Low-Temperature Entrainer Process of Fig. 9-5[a]

Stream	Flow rate (pph)	H (Btu/lb)	S (Btu/lb°R)	H (Btu/h)	T_0S (Btu/h)	$H - T_0S$ (Btu/h)
Incoming streams						
20-psig saturated						
steam	56,800	1167.1	1.687	6.629×10^7	5.079×10^7	1.55×10^7
Water in feed	5,992	77.91	0.147	466,837	466,837	0
						$A_{in} = 1.55 \times 10^7$
Outgoing streams						
20-psig condensate	56,800	227.9	0.381	1.294×10^7	1.147×10^7	0.147×10^7
H_2O at 168°F	5,988	135.9	0.244	813,769	774,268	39,401
						$A_{out} = 1.499 \times 10^6$

[a] From Reference 1.

The data is presented in Table 9-5, and the results are as follows:

$$\Delta A = -W = 1.499 \times 10^6 \text{ Btu/h} - 1.55 \times 10^7 \text{ Btu/h},$$

$$W_{act} = 1.400 \times 10^7 \text{ Btu/h},$$

$$\text{efficiency} = (0.623 \times 10^6)/(1.4 \times 10^7) = 0.045 \quad \text{(i.e., 4.5\%).}$$

For the high-temperature entrainer process of Fig. 9-6, the same assumptions are made, and the data are presented in Table 9-6. The results are as follows:

$$\Delta A_{min} = 0.623 \times 10^6,$$

$$W_{act} = 1.495 \times 10^7 \text{ Btu/h},$$

$$\text{efficiency} = (0.623 \times 10^6)/(1.495 \times 10^7) = 0.0417 \quad \text{(i.e., 4.2\%).}$$

From the availability changes calculated, we find two things. First, in common with the other fractionation process, the efficiency of this process is very low (about 4%). However, the location of the major work losses for this process are not known, because the thermodynamic analysis has not been carried out in enough detail to give that information. Second, the low-temperature entrainer process has a very small energy credit compared to the high-temperature process. Assuming all other factors are truly equal, there is little to differentiate these two processes on the basis of available energy consumption. Detailed estimates would also be required to determine what the capital cost differences between the two processes are. The low-temperature entrainer has a lower onsite investment, but it requires substantial cooling water backup from offsite. The high-temperature entrainer requires very little cooling water, because most of the heat is removed by the

Table 9-6

Availability Data for the High-Temperature Entrainer Process of Fig. 9-6

Stream	Flow rate (pph)	H (Btu/lb)	S (Btu/lb)	H (Btu/h)	$T_0 S$ (Btu/h)	$H - T_0 S$ (Btu/h)
Incoming streams						
175-psig saturated						
steam	93,500	1197.6	1.55	1.1198×10^8	7.673×10^7	3.525×10^7
Water at 110°F						0
					Total A_{in}	$= 3.205 \times 10^7$
Outgoing streams						
22-psig steam (net)	80,000	1158.1	1.724	9.263×10^7	7.306×10^7	1.957×10^7
22-psig condensate	2,500	201.3	0.3431	1.602×10^6	1.448×10^6	8.09×10^5
H_2O at 233°F	5,447	201.2	0.343			
175-psig condensate	11,700	334.3	0.5183	3.911×10^6	3.256×10^6	
					Total A_{out}	$= 2.03 \times 10^7$

generation of 20-psig steam. As a result, we can conclude that the choice between these two processes should be made on the basis of economics rather than available energy consumption.

It is instructive to demonstrate the impact of plant steam values on this choice. As discussed earlier, steam values are often assigned by accountants based either on a poor understanding of the value of steam or temporary steam balance problems. Table 9-7 shows an economic comparison of the energy effects of the two processes, with steam values priced on an availabil-

Table 9-7

Economic Comparison of the Low-Temperature and High-Temperature Entrainer Processes[a]

Cost or credit	Availability price[b] (dollars)	Enthalpy price[c] (dollars)
High-temperature entrainer		
Steam cost	3,740,000	3,740,000
Steam credit	2,182,000	2,974,000
Net cost	1,558,000	766,000
Low-temperature entrainer		
Steam cost per year	1,549,000	2,136,000
Incentive for low-temperature entrainer per year	9,000	−1,370,000

[a] Assuming the cost of 175-psig steam is $5/1000 lb.
[b] The availability price for 20-psig steam is $3.41/1000 lb for 8000 h/yr.
[c] The enthalpy price for 20-psig steam is $4.70/1000 lb for 8000 h/yr.

ity basis compared with those priced on an enthalpy basis. In both cases the 175-psig steam value was assumed to be $5/1000 lb, and the 20-psig steam value was derived by availability ratios in one case and enthalpy ratios in the other (compared to 175-psig steam). In the availability price column, the small advantage for the low-temperature entrainer process is reflected in the $9000/yr credit. A marked difference is seen if enthalpy prices are used, however. A $1.4 million/yr credit is shown for the high-temperature entrainer process. This misinformation actually swung the process selection in a real case. This was true even though the plant had assigned a value of $0 to 20-psig steam only a year earlier to reflect its short-term steam balance situation. Obviously, with free 20-psig steam an overwhelming advantage for the low-temperature entrainment process would have been calculated. Neither answer would have represented the true plant situation for the long term. As discussed earlier, developing appropriate prices for steam at various pressure levels is one of the most complex problems facing the energy conservation engineer. All values should be tested for reasonableness against a long-term set of thermoeconomic values for the plant in question to be sure that short-term effects do not lead to erroneous decisions.

D. Considering a Complex Process

The production of ethylene from the steam cracking of ethane feed is a complex process. Basically, the feed and steam diluent are heated to very high temperatures in a furnace reactor and immediately quenched in a waste heat boiler to keep the residence time short. The reactor effluents are subsequently cooled, compressed to the order of 500 psig, and then refrigerated, culminating in the cryogenic separation of methane and hydrogen from the C_2 and heavier compounds in the demethanizer and of ethane from ethylene in the product tower. Depending on the impurities in the ethane feed, various warm distillation towers are required to remove C_3 and heavier products. The entire process is shown schematically in Fig. 9-7. This process will be analyzed on an overall basis and then broken down into subsegments to demonstrate the approach for handling a very large and complex process.

Table 9-8 shows the ideal and lost work for the entire process, broken down into a number of process segments. The negative signs in the ideal work column indicate that theoretically it should be possible to recover work from these segments of the process rather than be necessary to apply work. As discussed in the section on methodology, it is important to draw the process boundaries carefully, particularly in such a complex problem as this.

Certain assumptions have been made in developing Table 9-8. For the

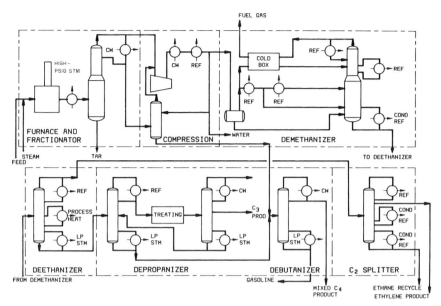

Fig. 9-7 Steam cracking process for the production of ethylene from ethane feed. From Reference 1.

most part, the ideal work calculations were made only for the processing segment of the plant. The utility sections were assumed to have no ideal work of their own, only supplying the work required by the process segments. Therefore, the utility segments have lost work but no ideal work, because the ideal work is included in the other process segments. It has also been assumed that it is not one of the objects of the process to heat cooling water, so the availability of the cooling water streams can be assumed to be zero. All data in Table 9-7 are given in standard horsepower. Overall, the process has an efficiency of about 22%. Note that the charge-gas compressor segment has an efficiency of about 60%, whereas those for the quench tower demethanizer and deethanizer are negative and that for the ethylene fractionator is about 14%.

At this stage of analysis the segments of the process that have the greatest inefficiencies can be identified. Not surprisingly, these include the cracking furnaces and the boiler and steam system, where large combustion components exist. However, there are three other segments of the process from which we should recover work but are forced to supply work instead. The ethylene fractionator–refrigeration system also has an efficiency that is lower than the overall process efficiency. We can therefore set up a priority list as follows:

Table 9-8

Breakdown of Work in the Thermodynamic Analysis of an Ethane Cracker (Fig. 9-7)

	Work[a] (hp)			
Section	Ideal work[b]	Lost work	Net actual work	Percentage of total lost work
Feed pretreatment	100	3,800	3,900	1.1
Cracking furnace and TLE	86,200	142,800	229,000	43.1
Quench tower	−12,300	12,800	500	3.9
Charge-gas compressor	19,200	13,300	32,500	4.0
Demethanizer	−200	6,100	5,900	1.8
Deethanizer	−1,700	4,100	2,400	1.2
Other towers	2,800	1,800	4,600	0.5
Ethylene fractionator and ethylene refrigeration system	1,900	11,900	13,800	3.7
Propane refrigeration system	—	8,500	8,500	2.6
Boiler and steam system	—	101,900	101,900	31.1
Electric power generation[c]	—	22,000	22,000	6.7
Total	96,000	329,000	−425,000	100.0

[a] Ideal work plus lost work equals actual work.

[b] Ideal work for a work-requiring process is positive, the ideal work for a work-producing process is negative.

[c] Electric power generation is assumed to be 33% efficient.

(1)   furnaces,
(2)   boiler and steam system,
(3)   quench tower,
(4)   ethylene fractionator – refrigeration,
(5)   demethanizer.

The next step is to break down some of these sections in more detail to pinpoint losses of availability more accurately. Table 9-9 provides this breakdown for some of the segments of interest. Two of these segments, the

Table 9-9

Detailed Breakdown of Lost Work in an Ethane Cracker (Fig. 9-7)

Section	Lost work (hp)
Cracking furnace and TLE steam	
Cracking furnace	
Radiant section	
Lost work of combustion	82,100
Lost work due to heat transfer and reaction	14,300

Table 9-9 (*continued*)

Section	Lost work (hp)
Convection section	
Lost work in flue gas	1,500
Lost work in hydrocarbon preheat	900
Lost work in mixed preheat (lower temperature)	6,500
Lost work in steam superheat	4,000
Lost work in mixed preheat (higher temperature)	14,900
Cracking furnace lost work, subtotal	124,200
TLE and steam generation	
Lost work in TLE	17,600
Lost work in BFW heaters	1,000
TLE and steam generation lost work, subtotal	18,600
Total, cracking furnace and TLE steam lost work	142,800
Quench tower	
Lost work in secondary TLE	4,100
Lost work in quench tower	8,300
Lost work in dilution steam generation	400
Total, quench tower lost work	12,800
Feed chilling and demethanizer	
Lost work due to feed chilling	1,860
Lost work due to off-gas expander	677
Lost work in demethanizer	3,244
Other	319
Total, feed chilling and demethanizer	6,100
C_2 fractionator and ethylene refrigeration system	
C_2 fractionator	4,500
Compressor	4,100
Condenser	1,000
Feed vaporizer	800
Product vaporizer	700
Valves	800
Total, C_2 fractionator and ethylene refrigeration system lost work	11,900
Boiler and steam system	
Boiler	
Lost work of combustion	29,800
Lost work due to heat transfer	28,400
Boiler lost work, subtotal	58,200
Deaerator	3,300
Turbines for charge-gas compressor and propane refrigeration compressor	29,700
Turbines driving other equipment	3,000
Atmospheric losses	7,700
Total, boiler and steam system lost work	101,900

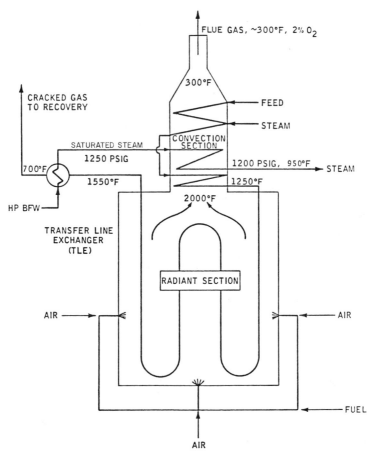

Fig. 9-8 Simplified process flow diagram for the furnace reactor of an ethane cracker. From Reference 1.

cracking furnaces and the demethanizer, will be explored to demonstrate how to derive some useful information about process improvement from thermodynamic analysis.

1. Furnace Section

Figure 9-8 is a schematic diagram of the process flow for the furnace reactor in an ethane cracker. Because of the complex kinetics of the reactions that take place, there are a number of restrictions imposed on the reactor system. These include the following:

(1)   limiting the process temperature entering the radiant section to about 1250°F;

(2)   allowing only a relatively short (less than 1 sec) residence time in the radiant section;

(3)   immediately quenching in the waste heat boiler at the reactor outlet;

(4)   allowing no appreciable reaction in the convection section;

(5)   creating a uniform distribution of heat input into the radiant section;

(6)   having the ability to deal with coking anywhere in the reactor;

(7)   obtaining a high furnace efficiency.

The cracking furnace section in Table 9-9 shows that the vast majority of lost work occurs in the combustion process itself. From previous discussions, this is not at all surprising. However, there are additional rather large components of lost work associated with the driving forces for heat transfer in both the radiant and convection sections of the furnace and in the transfer line exchanger (TLE) outside the furnace. From these data, what ideas might be generated for improving the thermodynamic efficiency of the furnace and the TLE segment of the process? As before, take some time now to list these ideas.

Clearly, our first objective should be to reduce the amount of fuel fired. However, because of the process restrictions listed earlier, there is little that can be done within the furnace itself without affecting process kinetics. Little work is lost in the flue gas, so a more efficient convection section would have little impact. Reducing the temperature differences in both the radiant and convection sections offers the potential for savings in lost work at the price of additional surface area, but how to accomplish this in such a constrained system is a problem. If we cannot change the process configuration to reduce the amount of fuel fired, what external effects might be manipulated to save fuel?

Several approaches are shown in the accompanying list.

Ideas for Furnace Efficiency Improvement

Add a topping cycle between the flue gas and the process
Preheat the air
Preheat the fuel
Reduce the stack temperature and flue-gas oxygen content
Increase the recovered steam pressure and the superheat temperature

---

One is a "classic" thermodynamic solution, namely, the interposing of a topping cycle between the flame and the process. This is theoretically possible, but again one must be careful not to upset appreciably the residence time or pressure drop in the reactor section. In addition, this is bound to require a significant capital cost. The second idea is to reduce the amount of fuel fired by preheating the combustion air and fuel, as is done in other process furnaces. This technology is not in common use in ethylene furnaces because of the requirement for many small burners to provide a uniform heat distribution in the radiant section. Other ideas that might be suggested include the generation of even higher-pressure steam in the TLEs and improving the degree of superheating of the steam that is generated. Both of these ideas would involve recovering more available energy from the reactor effluent than is presently practiced. There are clearly mechanical limitations to be dealt with in these areas that stem not only from furnace materials but also from machinery limitations.

Perhaps you generated other ideas. How do they fit into the constraints of the process and the objectives we are trying to accomplish?

The energy conservation engineer should be most interested in fuel and air preheating, because it attacks the major loss in the system. Introducing cold reactants to the combustion reaction is very similar in principle to using cold reactants to help control the exothermic heat of reaction in the process. Because the reaction occurs at a very high temperature, adding these cool streams absorbs heat through the very large temperature difference, and a large loss in work results.

Because of the ratio between air and fuel (20 lb air/lb fuel), about 400°F of air preheat amounts to a 10% saving in fuel. One can assume that this heat might be provided from another segment of the process to test the potential benefits of such a change on the furnace reactor being considered. An additional 0.5% fuel saving might be accomplished by preheating the fuel as well.

In the plant under consideration, a 10% fuel saving amounts to about 58 MBtu/h. This fuel saving also represents a 10% saving in the lost work of combustion. From Table 9-9 the lost work amounts to

$$W_{lost} = 82,100 \text{ hp} \times 2545 \text{ Btu/hp} \cdot \text{h} = 209 \text{ MBtu/h}.$$

The 10% saving is 21 MBtu/h in combustion alone.

The same amount of heat must be transferred in the radiant section, because the inlet and outlet temperature are fixed, as is the heat of reaction. Because the combustion-air temperature is higher, the flame temperature will be higher. Ten percent less air and fuel are required, resulting in 10% less flue gas. By stoichiometry, the flue gas will exit from the radiant section at about 2100°F rather than the original 2000°F. Thus, the temperature difference in the radiant section will be somewhat higher, resulting in an increase in lost work.

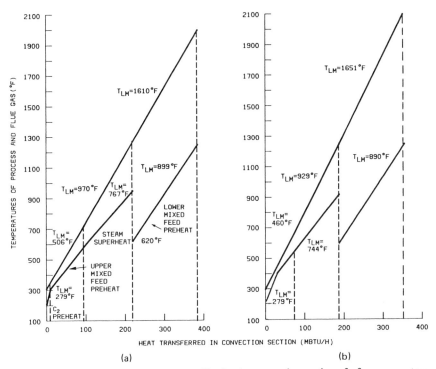

Fig. 9-9 Approximate temperature profiles for the convection section of a furnace reactor: (a) before air preheating; (b) after air preheating.

Changes in the lost work profile of the convection section can only be calculated by recalculating the temperature profiles for the reduced flue-gas rate. This has been approximated in Fig. 9-9. The only variable that can change is the steam superheat temperature, because feed preheat temperatures must be held constant, as is the steam flow rate. Thus, the duty on the convection section must be reallocated. By heat balance, the steam can only be superheated to 910°F, compared with 950°F originally. From the steam tables, it is seen that this represents a loss of product available energy of 7 MBtu/h. This loss is partially compensated for by less work being lost in this segment of the convection section because the temperature difference is less.

Changes in lost work profiles are calculated from changes in the temperature difference in each section. For example, in the mixed-feed preheat section, where steam and feed are heated to 1250°F just before the radiant section, the change in lost work is calculated from

$$W_{lost} = T_0 \, \Delta S - Q_{atm},$$

where $Q_{atm} = 0$. Therefore,

$$W_{lost} = (537)(\Delta S_{proc} - \Delta S_{fg})$$
$$= (537)\{\Delta S_{proc} - [Q/T_{LM}(\text{flue gas})]\}.$$

For the mixed-feed preheat before air preheating,

$$W_{lost} = (537)[(15.0 \times 10^4) - (165 \times 10^6)/(1610 + 460)]$$
$$= 37.7 \times 10^6 \text{ Btu/h.}$$

After air preheating,

$$W_{lost} = (537)[(15.0 \times 10^4) - (165 \times 10^6)/(1651 + 460)] = 38.6 \times 10^6.$$

The overall results are summarized in Table 9-10. The lost work saved in the furnaces is 26.8 MBtu/h. However, interactions in the steam system reduce this. As previously discussed, the high-pressure steam produced will contain 7 MBtu/h less available energy, and some of the waste heat used to preheat the air would probably have been used to generate medium-pressure steam. Rebalancing the steam power system will be required before the net impact on the amount of fuel consumed at the site is known precisely.

The purpose of this example was to demonstrate two points.

(1)   Even for situations in which few degrees of freedom exist, the problem of reducing the work lost in combustion (or other chemical reaction) can sometimes be attacked with conventional technology.

(2)   Even complex interactions resulting from a change in lost work can be identified and an approximation of their impact made in thermodynamic terms.

As noted in earlier examples, a great deal of conventional chemical engineering is involved in making any of these analyses. The lost work breakdown only identifies areas to explore, it does not guarantee that ideas for improvement will result.

2. Demethanizer System Analysis

Given the assumption that distillation is the only practical process step, some investigators support accepting the tower loss as inevitable to concentrate on the search for improvements elsewhere. In many cases this is a practical approach, but one should be careful to isolate the tower itself from the feed and other auxiliary systems. In Table 9-9 the feed chilling system and demethanizer tower are split into three subsections. Note that the biggest segment is the tower itself. The flash drums generating the liquid feeds to the tower are included with the chilling train losses. Figure 9-7

Table 9-10

Lost Work Results for the Convection Section of a Furnace Reactor[a]

Component	Lost work (MBtu/h)		Energy	Interactions	
	Base case	Air preheating			Capital
Combustion	209	188	−58 MBtu/h	0	
Radiant section	36.84	37	~0	~0	
Flue gas	3.8	3.5	~0	~0	
Superheater	10.1	7.1	911°F outlet results in a steam availability loss of 7 MBtu/h	Increased, lower ΔT	
Feed preheat					
High temperature	37.8	38.6	0	Decreased, higher ΔT	
Low temperature	19.2	15.3	0	Increased, lower ΔT	
Transfer line exchanger and boiler feed water	47.3	47.3	0	0	
Total	363.8	336.8			
Air preheat	—	—	Reactor or other waste heat is upgraded to fuel value from medium-pressure steam, lose medium-pressure steam availability		Probably increased because of lower coefficients

[a] From Reference 1.

shows a typical tower–feed system inside the dashed boundary labeled "Demethanizer."

The purpose of the demethanizer in an ethylene plant is to remove C_1 and lighter compounds overhead, concentrating C_2 and heavier compounds in the bottoms. Because each of the feeds represents a partial condensation of liquid from a H_2-rich stream, the liquids separated in the flash drums are subcooled relative to the dew points of the pure liquid mixtures. The net result is that, although the *compositions* of the feed streams match well the *compositions* of the liquids on the respective feed trays, the temperatures of some of the streams are significantly different than the tray temperatures, which are at equilibrium with the compositions.

A somewhat simplified flow diagram of the demethanizer tower system is shown in Fig. 9-10 for a 1 billion lb/yr steam cracker. The tower is fairly typical of a modern demethanizer-first ethylene recovery system. The tower operates at 530 psia. There are four feeds and an intercondenser to move some of the overhead duty at the −115°F refrigerant level rather than the −147°F level. An expander is provided on the overhead vapor to recover very cold refrigerant for the cold-box heat integration system.

The second law of thermodynamics teaches that mixing is inefficient. Here we are mixing streams of different temperatures, having overcooled some feeds to achieve condensation of liquids from a noncondensable-containing stream. The net result is typified by the third feed, which enters the tower at −143°F, when the temperature in equilibrium with the feed composition is about −96°F. Because all of the extra cooling is supplied by various levels of ethylene refrigeration, this could be a significant inefficiency.

The subcooling of the third feed is not inherent in the fractionation step (for which we have no practical alternative). A simple reheating system for the third feed, adding two new exchangers, might be proposed to save refrigerant (see Fig. 9-11). In this proposal the liquid from the flash drum MD-3 is reheated in the new exchangers to −95°F, reducing the load on the refrigerated coolers by 6 MBtu/h; 3 MBtu/h at the −115°F level and 3 MBtu/h at the −147°F level. However, to maintain the required separation in the tower, the refrigerant duties in the overhead condenser and intercondenser are increased by 0.4 MBtu/h (at −147°F and 1.4 MBtu/h at −115°F, respectively.

Because all of the refrigerant is normally provided in a C_3–C_2 cascade refrigeration system, savings appear as lower compressor horsepower requirements in both the propane or propylene (C_3) and ethylene (C_2) compressors. Thus, savings appear directly as work in this case and are usually cascaded back through the steam system to the boiler, as depicted in Fig. 9-1. The net savings for this case were about 2200 hp out of 42,000 hp base, or

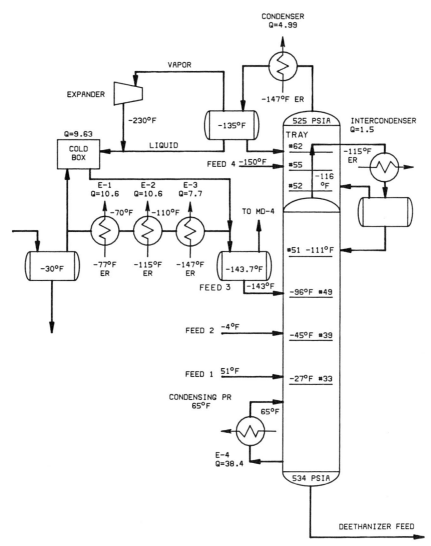

Fig. 9-10 Base-case flow sequence for a demethanizer tower system. All exchanger heat duties are in million Btu per hour. ER, ethylene refrigeration; PR, propane refrigeration.

about 5%. This is very significant and actually provides a debottlenecking capability on plant capacity in the summer. A fast economic evaluation of retrofitting these exchangers to the plant in question showed a 1-yr payout. It is not all clear that this is the optimum solution either.

Clearly, this reheating system could have been included in the original

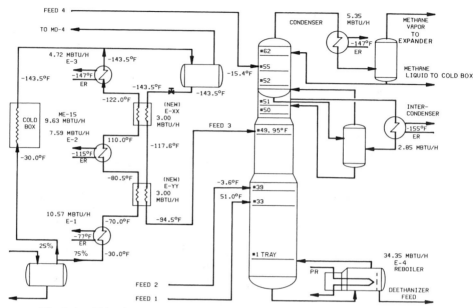

Fig. 9-11 Third-feed reheating process for a demethanizer tower system. ER, ethylene refrigeration; PR, propane refrigeration.

design of the plant if the designer had been more astute or had checked the design by thermodynamic analysis before issuing it. Omissions of this type spur development of design systems such as those labeled "process synthesis" and "thermoeconomics," about which more will be said in Chapter 11. The immediate economic impact of failing to eliminate this inefficiency is to increase the capital cost of the plant as well as lower its energy efficiency. The extra cost of compressors, steam systems, etc., more than offsets any savings on heat exchangers and piping.

In summary, note again that once the source of inefficiency had been identified, conventional process engineering was the only skill needed to define an improvement. Very little imagination was required to conceive the reheating system shown in Fig. 9-11, yet a major saving was identified.

III. Accepting "Inevitable" Inefficiencies

A. Principles

Denbigh[2] showed that separating the "inevitable" available energy loss (lost work) of a chemical reaction itself can sometimes be of value in focusing the search for energy efficiency improvements on those areas most likely to be improved with conventional technology. Basically, Denbigh derives the equation

$$W_{tot} = T_0 \, \Delta S + \Delta A, \tag{9-5}$$

or in words, the total work required for any process equals the lost work (T_0 times the entropy production of the process) plus the change in the available energy of the process materials.

The change in available energy is determined by the initial and final states of the materials and the temperature and pressure of the environment. This constitutes the "inevitable" work requirement, whereas the process engineer can make ΔS as small as economically feasible. Work can be extracted from processes in which ΔA is negative and larger than $T_0 \, \Delta S$.

Denbigh applies his analysis to the production of 60% HNO_3 from NH_3 at 20°C, with air and water at 15°C. T_0 was taken to be 288 K and P_0 to be 1 atm. The reactions are

(1) $\qquad NH_3 + 9.6 \text{ air } \xrightarrow[\text{1 atm}]{850°C} NO + 1.5H_2O(g) + 0.75O_2 + 7.6N_2,$

(2) $\qquad 1.34H_2O(l) + NO + 1.5H_2O(g) + 0.75O_2 + 7.6N_2$
$\qquad \rightarrow HNO_3 + 2.34H_2O + 7.6N_2.$

Reaction (1) occurs over a platinum catalyst and reaction (2) in an absorption tower. The 1.34 mol liquid water is simply that needed to dilute the product to 60% HNO_3. Under these conditions the available energy change for the reaction is negative (exothermic):

$$\Delta A = -74.5 \text{ kcal/g} \cdot \text{mol } HNO_3.$$

Thus, in a completely reversible process, this amount of work might be *recovered* from the reaction.

A typical practical process of the time (1956) *required* 4.1 kg·cal electricity/g·mol HNO_3 and *produced* 31.9 kcal 50-psig steam/mol HNO_3. Thus, heat (not work) is recovered from the process. From the steam tables, the work equivalent of the recovered steam can be calculated to be 8.8 kcal/g·mol (from $W = Q + T_0 \, \Delta S$).

The net work recovered is 8.8 kcal produced minus the 4.1 kcal requirement, or 4.7 kcal/g·mol HNO_3. This leaves a second law efficiency for the process of about 6% (4.7/74.5) versus the standard of a perfectly reversible process.

If one chooses to accept current reactor technology with its inherent irreversibilities as a standard, the result is a different answer and a different focus. Denbigh assumes that the entire process is carried out reversibly except for the two reactions, which occur at commercial conditions and efficiencies. The new total potential for work recovery in the process will be reduced by the "inevitable" lost work in the two reactions at prescribed conditions. The entropy production for each reaction is

$$\Delta S_R = \Delta S_{sys} + \Delta S_{env}. \tag{9-6}$$

Because no work is done by the reactants, ΔS_{env} is calculated from Q/T, which is $\Delta H_R/T$. Thus,

$$\Delta S_R = \Delta S_{sys} - \Delta H/T, \tag{9-7}$$

or

$$\Delta S_R = -\Delta G/T. \tag{9-8}$$

The lost work is therefore

$$W_{lost} = T_0 \Delta S_R \quad \text{or} \quad W_{lost} = -T_0 \Delta G/T. \tag{9-9}$$

For the reactions described earlier, standard data for the Gibbs free energy of reaction at 25°C and heat capacity data can be used to calculate ΔG at 850°C for the oxidation reaction as -78 kcal/g·mol NO and ΔG at 25°C for the absorption reaction as -12.7 kcal/g·mol HNO_3. The "inevitable" lost work is

$$W_{lost} = \sum -T_0 \Delta G/T = (288/1123)(78) + (288/298)(12.7)$$
$$= 32.3 \text{ kcal/g·mol } HNO_3.$$

The practical work that might be recovered (excluding the reaction losses) is thus

$$74.5 - 32.3 = 42.2 \text{ kcal/g·mol } HNO_3.$$

The "practical efficiency" of the process is thus 4.7/42.2 or 11%, still very low. But now we can focus on the facets of the process that might be improved by conventional technology. These include

(1) pressure drop through the system, both gas and liquid, which accounts for the 4.1 kcal electrical requirement;

(2) recovery of steam at 50-psig (138°C), when the reaction temperature is 850°C;

(3) incomplete recovery of reaction heat.

In addition, there is the major opportunity to carry out the reactions closer to reversible conditions. However, we have assumed these losses are "inevitable" with today's technology. For processes in which reaction losses tend to divert energy efficiency improvement studies from more productive areas, the Denhigh approach may be profitable. Unfortunately, efficiencies calculated by this method will change as each investigator defines his own

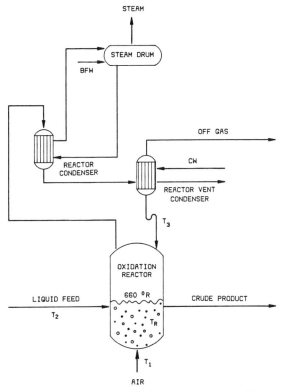

Fig. 9-12 Flow diagram for a process reactor system involving liquid-phase air oxidation.

"inevitable" lost work of reaction. The reader is cautioned to define his base case well in this approach.

The approach of isolating "inherent irreversibilities" has merit in certain cases. In setting up a system analysis, one can decide to accept certain irreversibilities as inevitable (at least with present technology) and concentrate on those inefficiencies that are less constrained.

Townsend[3] discusses an example of a process reactor system involving liquid-phase air oxidation. A flow diagram for this system is shown in Fig. 9-12. Air and liquid feed are introduced into the reaction vessel continuously, with the exothermic heat of reaction removed in a reactor solvent condenser, generating steam. The effluent from the solvent condenser flows through a vent condenser, in which cooling water is used to condense the remaining reactor solvent. Nitrogen and other inerts are vented from this vent condenser. A breakdown of the lost work in the system, calculated as exergy loss, shows that the reactor system represents by far the major loss in the process (57.64%), although losses in the solvent dehydration section (13.05%) and reactor air supply system (5.24%) are also significant.

What is the practical potential for improving the efficiency of the reactor system itself? As discussed in Chapter 7, the free energy change of a reaction at constant temperature is equal to the change in available energy of the reaction. For the reaction in question, the free energy change is -77.2 MW, the reaction being exothermic. Townsend also shows that the available energy balance (exergy balance) across the reactor system shows a change of -28.5 MW. Under ideal circumstances the entire 77.2 MW could have been recovered. However, this would have required a reactor system capable of recovering work directly from the reaction itself, similar to a fuel cell device. Such a reactor system does not yet exist, limiting practical energy recovery to the ability to extract heat from the reaction. If the actual change in available energy across the reactor system, namely, 28.5 MW, is subtracted from the free energy change of the reaction (77.2 MW), the lost work of the reactor system is 48.7 MW. Following Denbigh's analogy,

$$\Delta A(\text{inevitable}) = -(T_0/T_R)\,\Delta G. \qquad (9\text{-}9)$$

Given the reactor temperature at $660°F$ and $T_0 = 537°R$, the inevitable available energy change of the reaction is 39.6 MW. This implies that

$$\Delta A(\text{avoidable}) = 48.7 - 39.6 = 9.1 \text{ MW}.$$

These calculations show that approximately 80% of the total lost work of the reactor is attributable to the fact that we cannot extract work directly from the reaction. A further breakdown of exergy losses in the reactor system is given in Table 9-11. If we correct the table to eliminate the "inevitable" reactor losses, we obtain a somewhat different picture of the significance of reactor losses that might be subject to improvement without

Table 9-11

Exergy Losses in the Reactor System of Fig. 9-12

Component	Exergy loss (irreversible) (MW)	Percentage of total
Reactor steam drum	0.08	0.1
Reactor condensers	3.95	6.8
Reactor vent condensers	4.95	8.6
Oxidation reactors	48.66	84.5
Total	57.64	100

Table 9-12

Exergy Losses in the Reactor System of Fig. 9-12 with
Inevitable Reactor Loss Discounted

Component	Exergy loss (irreversible) (MW)	Percentage of total
Reactor steam drum	0.08	0.4
Reactor condensers	3.95	21.8
Reactor vent condensers	4.95	27.8
Oxidation reactors	9.12	50.0
Total	18.10	100

major changes in technology. This is shown in Table 9-12. Here the "preventable" losses in the reactor system are approximately one-third of the total loss of the reactor, and although the reactors are still very significant, they are much more nearly at the same order of magnitude as the reactor condensers. With these figures the sources of the avoidable losses can be explored without complication by large inevitable losses.

The heat of reaction in this type of reactor is removed by solvent evaporation into an inert (nitrogen plus unreacted oxygen) gas stream. In addition, part of the reactor heat is degraded internally in the reactor by heating up the reactor air, the feed, and the reflux, because the temperatures of all of these are lower than the reactor temperature. Thus, more steam could be raised in the reactor condenser if one, the temperatures of the feed materials were higher, causing more solvent evaporation, and two, it were not necessary to subcool the effluent from the reactor condenser to such an extent in the vent condenser to prevent losses of solvent with the off-gas.

A list of ideas for reducing the avoidable lost work in this reactor system can be generated. These include the following:

(1) investigate further preheating of reactor feeds;
(2) reduce the inert gas flow into the reactor, which causes subcooling of the reflux;
(3) eliminate the solvent system altogether and explore the possibility of direct removal of heat from the reactor.

The obvious limit in reducing the inerts used in the reaction is to switch from air to oxygen.

It is probably true that an astute process engineer would arrive at virtually the same list of process improvement ideas regardless of the route chosen. However, for the less experienced, a systematic approach tends to focus thinking on the areas for which logical improvements can be conceived.

The range of technology suggested for improving this process is also of interest. Further preheating of the feed streams requires little in technological development, but may impose more severe control problems on the reaction itself. Reducing the inert content of the air stream implies either concentrating the oxygen from the air or reducing the excess oxygen in the reactor system. Within limits, either can be accomplished with proven technology. However, attempting the direct removal of heat from the reactor would probably involve a significant research project. It seems reasonable to assume that if the characteristics of the reaction mass were suited to the direct removal of heat, this would have been done in the first place. Given some additional data about the temperatures and flow rates in the reactor system, the incentives for this development could easily be calculated. The combination of these ideas into a practical result is also worth consideration. Would it be feasible to use the reactor feed stream as part of the coolant in the vent condenser? Would it be feasible to generate power in the vent condenser, by an organic Rankine cycle, to provide some of the horsepower for air compression? Can the heat rejected from the vent condenser be combined with heat rejected elsewhere in the process to make a more economical system for power recovery? By cascading creative thinking in this manner and being cognizant of potential interactions elsewhere in the process, a number of practical systems can be defined for economic analysis. The combination of comparable lost work sources into a single recovery system is a strategy more likely to produce an economic result than the individual handling of each lost work source. Again, the strategy used should be sensitive to the interactions of the system rather than being victimized by them.

IV. Optimization through Lost Work Analysis

In some cases it may be possible to simplify the energetics of a process into relationships that make possible optimization studies based on a few simple variables. An example of this approach was provided by Gyftopoulos and Bennedict.[4] The authors investigated a practical air separation process of some years ago, which is illustrated in Fig. 9-13. A thermodynamic analysis of the process in terms of the irreversible entropy change was carried out. The results are shown in Table 9-13. For the practical process $T_0 \Delta S$ was approximately 23.5 MBtu/h, as opposed to the theoretical work requirement of 4.6 MBtu/h. Thus, the efficiency of the process can be calculated from the following:

$$\text{efficiency} = \text{theoretical work/actual work} = 16.4\%. \qquad (9\text{-}10)$$

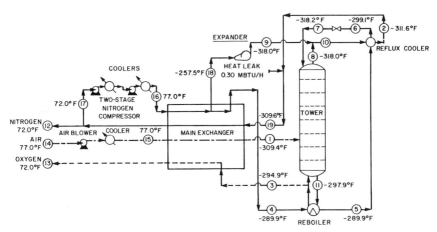

Fig. 9-13 A practical air separation process. From Reference 6.

The authors conceived an ideal process in which the irreversibilities associated with factors that could be alleviated with increased investment were minimized (see Fig. 9-14). The resulting thermodynamic analysis is shown under the heading "Ideal process" in Table 9-13. The irreversibilities

Table 9-13

Breakdown of Entropy Production in the Air Separation Process of
Fig. 9-13[a]

Component	Ideal process		Practical process	
	Work input (10^6 Btu/h)	$T_0 \Delta S_{irr}$ (10^6 Btu/h)	Work input (10^6 Btu/h)	$T_0 \Delta S_{irr}$ (10^6 Btu/h)
Air blower	—		5.11	0.988
Air cooler	—			0.268
N_2 compressor	7.72	—	23.88	5.109
N_2 coolers	—			2.967
Main exchanger		0.927	−0.87	6.409
Expander	—			0.671
Tower		2.222		2.886
Reboiler		0.089		1.583
Reflux cooler		0.161		1.448
Valve		0.101		0.427
Heat leak	—			0.787
Total	7.72	3.50	28.12	23.54
Theoretical work $= H - T_0 \Delta S$		4.21		4.61
Predicted work (10^6 Btu/h)		7.71		28.15

[a] The feed rate is 10,000 mol air/h. From Reference 6.

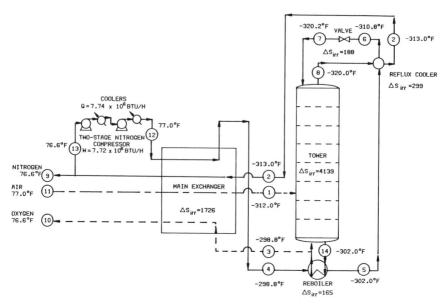

Fig. 9-14   An ideal air separation process. From Reference 6.

of the tower itself were largely accepted, but those of the machinery and heat exchangers were reduced by unconstrained investment in heat exchange surface and machinery. The limiting case for process efficiency was thus defined, and the analysis served to isolate the major sources of irreversibility to the nitrogen compressor and the main heat exchanger. After reaching this conclusion, the authors explored the optimum design point. They developed equations that related the area of the heat exchanger and its irreversible entropy change to two controllable design variables: the pressure drop across the heat exchanger (which affected the compressor horsepower and the heat transfer area) and the temperature driving force at the pinch in the exchanger. These equations are as follows:

$$\text{exchanger surface} = (1.886 \times 10^6)\, \Delta P^{-0.4}\, \Delta T^{-1.4}, \tag{9-11}$$

$$\text{compressor power} = 0.27\, \Delta S = (0.27)[3165\, \Delta P + 776\, \Delta T]. \tag{9-12}$$

The net result is that optimization of the process could be quickly accomplished, because the key variables that affect energy consumption and capital cost were isolated in these simple equations. Given some assumptions about the cost of energy and capital in 1949, the authors proceeded to optimize the process as shown in Fig. 9-15. Total costs for capital and energy were related to the pressure drop at several exchanger temperature differences. Their studies showed that the optimum pressure drop across the heat exchanger was much lower than that normally designed.

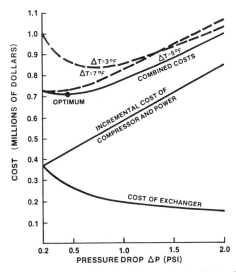

Fig. 9-15 Cost balance for the main heat exchanger in the ideal air separation process. Exchanger surface area $A = 1.886 \times 10^6 \Delta P^{-0.4} \Delta T^{-1.4}$; $\Delta S_{irr} = 3165 \Delta P + 776 \Delta T$; power (kilowatts) $= 0.27 \Delta S_{irr}$. From Reference 4.

We shall return to this example in Chapter 11 to discuss further the impact of reduced entropy production on the capital costs of facilities. The key issue here is that, having related the energetics of the process and the capital cost to two simple design variables, optimization was carried out simply and quickly. Clearly, sophisticated analysis was required to boil down a practical process to two variables. However, the result provides the design engineer with clearly understandable guides for trading capital costs and energy efficiency. This type of process representation would also be a great asset to process developers and researchers as they bring new reactions and process routes from the laboratory into production.

V. Research Guidance

A. Introduction

As alluded to in the previous section, a fundamental understanding of the energetics of a process can often be used in the early stages of new process development. In fact, the most cynical of process designers are willing to acknowledge the value of a concise understanding of process energetics, even though they doubt the ability of a simple system to provide a useful process design tool. In the early stages of process development, thermodynamic analysis can be done quite simply. Only concepts need be dealt with, not the details of hardware. Many investigators have found that processes that are not fully proven can be studied profitably by the preparation of a flow chart and the determination of the location and magnitude of the irreversibilities. The results of this type of flowchart analysis have often been used to suggest appropriate experimental work. It may even be argued that the most effective way to develop a process is to cause continuous interaction between the preliminary design work and the laboratory experimental work using thermodynamic analysis.

In the examples to be discussed, an attempt will be made to demonstrate how some very simple, even qualitative, thermodynamic analysis can help identify potentially profitable research directions. The objective will be to give readers some understanding of how to examine their own process problems with an eye toward practical directions for improvement.

B. Polymerization Process

In general, it seems simplest to demonstrate concepts with examples in which work is directly required to cause or permit a desired change. The relation to available energy input is clear, and incentives are easily calculated. A simplified version of such a process, which involves the polymerization of a hydrocarbon in a hydrocarbon diluent to produce a slurry of polymer is diagrammed in Fig. 9-16. The reaction operates at $-140°F$ for two reasons:

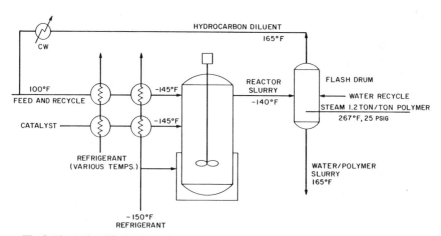

Fig. 9-16 A simplified polymerization process. The total refrigerant power is equivalent to 1050 hp/ton of reactor polymer.

(1)   with current catalyst technology, the desired products are produced only in this temperature range;

(2)   as the product slurry is warmed up in the hydrocarbon diluent, the particles tend to stick together and form lumps that plug the system.

As a result, the diluent is immediately changed from hydrocarbon to water by sparging with 25-psig steam. This vaporizes the hydrocarbon diluent and provides violent agitation to produce a water–polymer slurry suitable for further processing.

Reactor cooling is provided by two mechanisms: slight subcooling of feed and diluent and refrigeration in the jacket of the reactor. Total refrigeration horsepower requirements are 1050 hp (2.7 MBtu) per ton of reactor polymer. Low-pressure steam is added at a ratio of about 1.2 ton (2.4 MBtu) per ton of reactor polymer, representing an additional available energy input (at $T_0 = 77°F$) calculated from

$$A = (H - H_0) - T_0(S - S_0)$$
$$= (1169.7 - 45.02) - (537)(1.6763 - 0.09876)$$
$$= 271.5 \text{ Btu/lb steam} = 0.7 \text{ MBtu/ton reactor polymer.}$$

The vast majority of available energy consumption is related to the refrigeration, even though the enthalpy input of the steam is comparable in terms of Btu's.

Additional formal thermodynamic analysis is not required to obtain an overall view of the incentives for process improvement. The curves presented in Chapter 8 showing the variation of horsepower requirement per Btu of refrigeration as a function of the temperature at which the refrigeration is required indicate the potential for improvement. In this process various levels of refrigeration are used to chill the feed and the recycle diluent, but all of the heat of reaction and agitator power must be removed at the $-150°F$ level. The ability to run the reaction at temperatures closer to ambient would have two benefits:

(1)   there would be less duty required of the refrigeration system;

(2)   there would be less horsepower required per Btu of refrigeration duty.

The total incentive for a process at ambient (or higher) temperature is 1000 hp/ton polymer.

The conclusion is obvious, even trivial. Research should be done on a catalyst that provides the desired product at a higher temperature. If this were not a formidable task, it would have already been accomplished. However, several additional alternatives are worth considering.

The first alternative involves product as well as process research. Are there applications and/or formulations in which the product produced at higher temperature is of value? This might mean breaking into a new product market, or creating a different formulation of the existing product, but the incentives for accomplishing this objective can quickly be calculated from the horsepower–refrigeration charts and the number of tons of polymer produced per year.

Another alternative is to explore the possibility of heat integration (in this case refrigeration recovery) in the existing process. The reaction could remain at $-140°F$, but available energy recovery through a feed effluent–exchanger would also represent a sizable process improvement. To accomplish this two developments would be required: first, a heat exchange device that would provide a short enough residence time and sufficient turbulence to keep the slurry in suspension; second, some sort of slurry stabilization additive to prevent coagulation of the solids during the heat transfer process. A third alternative would be to find a new diluent that would allow a more stable slurry at higher temperatures. It might also obviate the need to switch slurry diluents in the flash drum, vaporize at lower temperatures, have a lower specific heat, etc., to minimize further the refrigeration requirements at any temperature.

What the author hopes to demonstrate is that relatively qualitative analysis from the second law point of view can identify the prime focus for research. The generation of ideas is still the challenge for the researcher, but there are well-founded priorities for guidance.

A more detailed thermodynamic analysis of this process might also prove profitable. In such a work-sensitive system, it would be well to know the optimum temperature differences for heat transfer, the impact of losses through insulation, the optimum agitator horsepower, and the details of the work losses in the flash drum. All these factors tend to involve conventional technology, not laboratory research. Thus, the analysis would be similar to those examples discussed earlier in this chapter.

C. Synthetic Natural Gas Process

Let us turn now to the discussion of the overall analysis of a more complex process. In 1978 Gaggioli, Rodriguez, and Wepfer analyzed the Synthane process for the Department of Energy.[5] The process consisted of the assembly of many small steps, all using presumably proven process technology, into a hopefully economic whole. The analysis was carried out in an early stage of the process evaluation effort. Its purpose was to verify that the best available technology was being used by the various engineering contractors

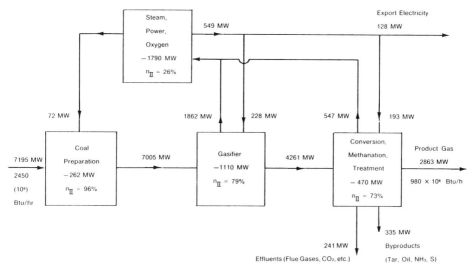

Fig. 9-17 Conceptual design for a Synthane process, including available energy flows. Negative numbers represent available energy consumptions.

working on the evaluation and to suggest areas for the study of improvements. The study was based on preliminary flow diagrams supplied by the contractor, various vendors, and in some cases, only very limited data on proprietary process steps.

A simplified conceptual design together with available energy flows is shown in Fig. 9-17. Various process steps have been grouped together in the four process areas shown. Through steam reforming and gasification with oxygen, coal is converted into a synthetic natural gas, some heavy hydrocarbon and chemical byproducts, and CO_2 and other flue gases. The available energy consumptions and the second law efficiencies (η_{II}) of each block are also shown on the flow diagram. These overall balances were obtained through a more detailed analysis of each process unit. Some of these will be discussed later, but an immediate observation can be made.

In a modern large-scale utility block, the second law efficiency should be much higher than 26%. About 30–35% would be typical of industrial utility systems designed with today's energy costs. As a result, an immediate suggestion is to analyze the utility block for efficiency improvements, because the conceptual design is not even up to par with today's proven technology.

The gasification section of the plant is seen to be the major consumer of available energy next to the utilities. This section consists of the gasification unit itself and the raw-gas quench unit. A detailed breakdown of the

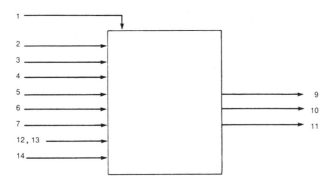

Inputs		Outputs	
1. Coal from Unit 14	5966.56	9. Gas to Unit 16	3661.17
2. CO_2 with coal	53.18	10. Steam and char to Unit 30	1599.53
3. Oxygen from Unit 32	30.00	11. Hopper vent gas	1.00
4. 1450-psig steam to gasifier	142.0	Total	5261.70
5. Purge gas from Unit 23	2.68		
6. 1450-psig steam to eductor	14.4		
7. 1450-psig atomizing steam	4.87		
12, 13. Process water from Unit 17	24.23		
14. 150-psig steam to char lock hoppers	1.67		
15. Electricity	1.0		
Total	6240.59		

Fig. 9-18   Availability flows for gasification in the Synthane process. All flows are in million kilocalories per hour. The second law efficiency is 84.33%, and the availability consumption is 978.8 million kcal/h.

available energy flows into the gasification unit is shown in Fig. 9-18. In the gasifier coal is partially combusted with oxygen to form the crude synthetic gas. Of nearly 6000 units of available energy fed with the coal, some 3660 are recovered in the crude gas, with another 1600 recovered in the steam and char fed to the utility system. Thus, about 60% of the available energy in the feed is recovered in usable form for other process needs and the production of export power. The available energy consumption in the gasifier is 978 units, leading to an overall efficiency η_{II} (recovered A/input A) of 84.3%.

This large available energy consumption is caused by the large irreversibilities of the combustion reactions, i.e., the chemical potentials of the reactants are much larger than those of the products. The rate of available energy consumption in this reactor could be reduced by raising the chemical potential of the products (e.g., by raising the pressure) or by reducing the chemical potential of the reactants. Gaggioli *et al.* point out that research on concentration cells, which would be similar to those used for monitoring the composition of combustion gas in today's utility plants, might provide techniques to extract work from the reactants as they diffuse into the burning process. During the combustion process each reactant diffuses through the other gases to the flame at a rate that is very roughly proportional to the gradient in its partial pressure. In the flame itself, the partial pressure of reactants is very low because of the speed of the reaction. The larger the difference in the partial pressure between each reactant and the flame, the more rapid the diffusion of the reactant to the flame. The larger the gradient in partial pressure, the larger the consumption of available energy in the diffusion process (just as the larger the temperature driving force in the heat exchanger, the greater the loss of available energy in the heat exchange process). By allowing a gaseous reactant to be drawn into the flame through a concentration cell that would yield an electric output while dropping the partial pressure of the reactant, a significant portion of the available energy that is usually consumed by diffusion would be obtained electrically. The partial pressure of the reactant in the reactor would now be lower and the diffusion to the flame slower. Consequently, the reaction would not proceed so rapidly and the reactor would need to be larger. If only 15% of the consumption in this section were saved through such a device, the byproduct power output of the overall process would be increased by about 160 MW, or more than double the current export power production.

Obviously, such concentration cells would be applicable to all combustion processes, not only to the gasifier of the Synthane process. This research suggestion represents something of the classic "wish list" of a thermodynamicist, namely, a more reversible approach to the combustion process. Yet concentration cells are in use as meters today, developing electric signals that are interpreted in terms of the oxygen or carbon monoxide concentration in flue gases. The authors' conclusion is that with some serious research, this "wish" may not be too far from fruition. Perhaps some of the "inevitable" losses in reactions can be reduced after all.

Alternatively, one might provide oxygen to the combustion process not in its molecular form, but rather in some partially oxidized (lower available energy) state. If the reactions that produce this partially oxidized state permit the recovery of work, then a two-stage combustion process might lead to some of the savings identified earlier. A chemical species which in the oxide form reacts with carbon to form CO_2 and a carbide, but later reacts with water to regenerate the oxide form is thermodynamically possible. This cyclic oxidation and reduction might provide better opportunities for recovering work than the standard combustion process.

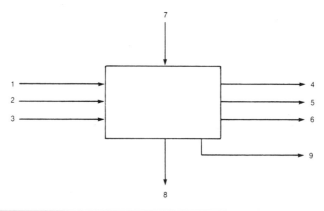

Inputs		Outputs	
1. Product gas from Unit 17	2885.98	4. Product gas to Unit 19	2642.05
2. Steam from Unit 30	13.50	5. Acid gases to Unit 21	38.82
3. Heat exchange from Unit 19	28.69	6. CO_2 to vent and to Unit 14	217.30
7. Shaft power	56.00	8. Electricity (net)	1.5
Total	2984.17	9. Light oil	98.3
		Total	2997.97

Fig. 9-19 Availability flows for acid gas in the Synthane process. All flows are in million kilocalories per hour. The nominal efficiency is 100.5%.

These examples clearly demonstrate that formulating research ideas from the data of thermodynamic analysis is not an "automatic" process. It still requires insight and other scientific knowledge. Yet the focus on the problem provided by the analysis helps.

Returning to other sections of the process, another important observation of the authors is that two of the proprietary processes included in the conceptual design of the overall system have interesting diversions from practical efficiency values. Although knowing almost nothing about these processes, the authors were able to identify potential problems and opportunities that justified further investigation. This would be a real asset to any purchaser of proprietary technology. One of these units is the acid gas removal unit shown schematically in Fig. 9-19. The interesting aspect of the analysis is that the second law efficiency of the unit is over 100%. Since this is thermodynamically impossible, either the data supplied are incomplete or incorrect or the process will not perform as claimed by its vendor. Clearly, further analysis is required to resolve this anomaly and ensure that the proprietary unit will meet its duty in the overall process.

As an aside to this discussion, it should be pointed out that it would not be all that unusual for erroneous data to be supplied by vendors of proprietary

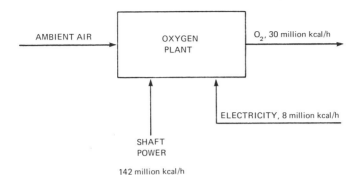

Fig. 9-20   Availability flows for the oxygen plant in the Synthane process.

processes. The author is familiar with at least one other case in which the contractor proposed a process with an inadequate power consumption. An overall thermodynamic analysis of the process showed that the refrigeration horsepower required in the second stage of a cascade refrigeration system to condense the refrigerant in the first stage was not included. When the corrections were made the proposed process was no longer economic.

The second observation of the authors about proprietary processes has to do with the oxygen plant. From the overall balance shown in Fig. 9-20, the oxygen unit is seen to be about 20% efficient. The output is about 30 units, whereas the consumption is twice what would be expected from current commercial units. Either the engineering contractor de-rated the oxygen plant vendor's proposal or secured a very low-capital version of proven technology with an exceedingly large debit in efficiency. Increasing the oxygen plant efficiency to a typical commercial design level of about 45% liberates an additional 97 MW of shaft power that could be added to the export electricity capability of the unit. This difference from current technology is thus very significant and needs to be explored further.

Summary

The objective of this chapter is to use a variety of examples to demonstrate how thermodynamic analysis might be used to suggest improvements in process sequences and hardware arrangements and to give guidance to research programs. Both simple and detailed analyses are shown to give the reader some feel for the mental processes involved. No attempt has been made to try to cover all types of processes or to develop a fool proof road map to creative thinking. Therein lies the dilemma of thermodynamic analysis: the engineer must still make the creative leap to the new process idea.

In spite of this, systematic use of the data provided by thermodynamic analysis can lead to insights helpful to the innovator in many ways. Both the areas in which to work and the potential magnitude of savings can be identified. Thus, priorities can be set for development efforts.

Practical applications of this approach have been found helpful in a number of areas, including

(1) research guidance,
(2) proposal evaluation,
(3) process development,
(4) process selection,
(5) heat exchange system analysis,
(6) process design updating,
(7) cogeneration systems development.

The relation between thermodynamics and economics again arises, and it will be discussed in more detail in the next two chapters.

Problem: Phthalic Anhydride Process Improvement

A somewhat different thermodynamic analysis has been carried out for the oxidation of orthoxylene to phthalic anhydride. The process is shown in the flow diagram of Fig. 9-21, with the material balance for the process given in Table 9-14. Preheated orthoxylene is sprayed into a compressed air mixture at about 275°F and 35 psia. The mixture flows into a catalytic reactor in what resembles a very large heat exchanger. The tubes are filled with catalyst and the heat of reaction is removed by a molten salt bath circulating on the outside of the tubes. The salt circulates through a high-pressure steam generator, which removes the heat of reaction by generating 300-psig steam. The effluent from the reactor flows through a second waste heat boiler in which 140-psig steam is produced and then into a group of switch

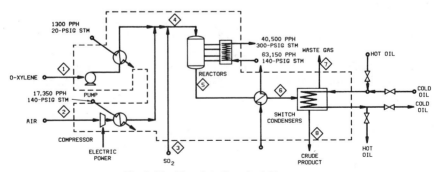

Fig. 9-21   The phthalic anhydride process.

Table 9-14

Material Balance for the Basic Phthalic Anhydride Process

Parameter	①	②	③	④	⑤	⑥	⑦	⑧
				Steam number				
Temperature (°F)	68	68	68	275	743	374	149	302
Pressure (psia)	15	15	40	35	30	30	18	15
N_2 (pph)	—	414,200	—	414,200	414,200	414,200	414,200	—
O_2 (pph)	—	125,100	—	125,100	102,676	102,676	102,676	—
H_2O (pph)	—	5,820	—	5,280	15,416	15,416	15,398	—
o-xylene (pph)	16,268	—	—	16,268	—	—	—	—
SO_2 (pph)	—	—	81	81	81	81	81	—
CO (pph)	—	—	—	—	2,420	2,420	2,420	—
CO_2 (pph)	—	—	—	—	9,276	9,276	9,276	—
Maleic anhydride (pph)	—	—	—	—	818	818	785	33
Byproduct acids (pph)	—	—	—	—	129	129	45	84
Residual (pph)	—	—	—	—	80	80	—	80
Phthalic anhydride (pph)	—	—	—	—	16,373	16,373	218	16,009
Phthalic acid (pph)	—	—	—	—	—	—	—	163
$H - T_0S$ (MBtu/h)	-5.94	-510.05	-0.21	—	—	-630.19	-613.45	-27.48

Table 9-15

Thermodynamic Analysis of the Basic Phthalic Anhydride Process

Component	A_{in} (MBtu/h)	A_{out} (MBtu/h)
o-xylene	−5.94	—
Air	−510.05	—
SO_2	−0.21	—
Low-pressure steam	0.36	—
Medium-pressure steam	6.33	23.03
High-pressure steam	—	16.32
Low-pressure condensate	—	0.03
Medium-pressure condensate and boiler feed water	3.75	1.03
High-pressure boiler feed water	3.33	—
Electricity	38.8	—
Waste gas		−613.5
Crude product		−27.5
Total[a]	−463.6	−600.6

[a] $W = -\Delta A = -(-600.6 + 436.6) = 137$ MBtu/h.

condensers. Here a number of condensers are operated in parallel, each cooled by a cold oil stream. Phthalic anhydride crystallizes from the vapor mixture, fouling the heat transfer surface of the tubes. The waste gases, containing maleic anhydride, carbon dioxide, and heavier materials, flow out to a waste disposal system. As the crude phthalic anhydride builds up on the tubes of an operating condenser, more material is lost in the waste gas. The hardware arrangement allows for automatic switching of the product gas to another condenser and a product recovery step in which hot oil is piped through the condenser tubes in place of cold oil, causing the crude phthalic anhydride to melt off the tubes into a receiver. Several condensers are involved in each plant, and they are automatically cycled from hot oil to cold oil operation by means of a complex instrumentation system. The thermodynamic analysis, which uses a different sign convention than recommended in this text, is given in Table 9-15. Using this data and the flow diagram, make a list of ideas for improving the process. Estimate quickly the impact of a major improvement you suggest. Comment on where the available energy is no longer useful.

Solution

Idea lists are not unique; there are as many versions as there are idea generators. One example is shown in the accompanying list, based on three major strategies derived from the available energy data: recover

Ideas for Improving the Phthalic Anhydride Process

1. Recover steam at higher pressure from the reactor and waste heat boiler
2. Preheat the feed further to allow more high-pressure steam generation
3. Recycle waste gas
4. Reduce air flow
5. Reduce the average temperature of the waste gas to improve recovery and reduce waste processing

Table 9-16

Material Balance for the Alternate Phthalic Anhydride Process, with Less Air Per Pound of o-xylene[a]

Parameter	\(\diamondsuit\)1	\(\diamondsuit\)2	\(\diamondsuit\)3	\(\diamondsuit\)4	\(\diamondsuit\)5	\(\diamondsuit\)6	\(\diamondsuit\)7	\(\diamondsuit\)8
					Steam number			
N_2 (pph)	—	276,130	—	276,130	276,130	276,130	276,130	—
O_2 (pph)	—	83,400	—	83,400	60,976	60,976	60,976	—
H_2O (pph)	—	3,880	—	3,880	13,476	13,476	13,476	—
$H - T_0 S$ (MBtu/h)	−5.94	−340	−0.21	—	—	—	−421	−27.5

[a] This table shows only data that differs from that for the basic process; all other data are identical to that in Table 9-14.

Table 9-17

Thermodynamic Analysis of the Alternate Phthalic Anhydride Process
(Reduced Air Flow)

Component	A_{in} (MBtu/h)	A_{out} (MBtu/h)
o-xylene	−5.94	—
Air	−340.0	—
SO_2	−0.21	—
Low-pressure steam	0.36	—
Medium-pressure steam	4.18	15.0
High-pressure steam	—	16.32
Low-pressure condensate	—	0.03
Medium-pressure condensate and boiler feed water	2.53	0.7
High-pressure boiler feed water	3.33	—
Electricity	23.62	—
Waste gas	—	−421.0
Crude product	—	27.5
Total[a]	−312.1	−415.9

[a] $W = -\Delta A = -(415.9 + 312.1) = 103.8$ MBtu/h.

more available energy from the heat of reaction (ideas 1 and 2), reduce the ΔA of the reaction mass (ideas 3 and 4), and recover more product per unit of available energy expended. The first and third ideas are straightforward engineering suggestions. The second requires changes in reactor compositions. Obviously, reducing air flow saves compressor horsepower, but it also impacts on all downstream losses. A revised heat and material balance (prorated) is shown in Table 9-16 for an arbitrary one-third reduction in air flow. A new thermodynamic analysis (also prorated) for the major process change is shown in Table 9-17. The change in ΔA is reduced by twice the air compressor savings. Whether this process is safe or economical is yet to be determined. Note that once the idea has been generated and its impact assessed, the development process reverts to conventional process engineering procedures, i.e., checking of yields, catalyst life, controllability, and above all the safety aspects of the improved process.

Notes to Chapter 9

1. W. F. Kenney, "Thermodynamic Analysis for Improved Energy Efficiency." AIChE Continuing Education, New York, 1980.

2. K. G. Denbigh, The Second Law Efficiency of Chemical Processes. *Chemical Engineering Science* 6 (1), 1 (1956).

3. D. Townsend, Second Law Analysis in Practice. *The Chemical Engineer* Issue No. 396, 628 (1980).

4. E. Gyftopoulos and M. Bennedict, Air Separation Process. American Chemical Society Symposium Series 122, "Thermodynamics—Second Law Analysis," Washington, D.C., 1980.

5. R. Gaggioli, L. Rodriquez, and W. Wepfer, A Thermodynamic Analysis of the Synthane Process. U.S. Department of Energy Report COO-4589-1, November 1978.

6. E. Gyftopolous, L. J. Lazaridis, and T. F. Widmer, "Potential Fuel Effectiveness in Industry." Ballinger, Cambridge, Massachusetts, 1974.

10

Thermodynamics and Economics,
Part II: Capital – Cost Relationships

Background Information

The usual expectation in the process industries is that improved energy efficiency requires increased investment. Much of this expectation is derived from the "fix-up" nature of early energy conservation activities, from a lack of understanding of the complete plant energy system, and from a deep-seated conviction that nothing comes free. However, from the more fundamental viewpoint of the second law of thermodynamics, one might expect to be able to reduce both the capital investment of a plant and its inefficiency at the same time. This comes from the knowledge that the second law efficiency of most industrial processes is very low and from the feeling that if less fuel is burned on a given site, then surely less hardware needs to be installed to handle the products of combustion.

Various energy policy studies performed in the aftermath of the 1973 oil embargo showed that on a global scale the required investment for efficient facilities was less than that for a number of conventional processes and their associated energy supply systems. Gyftopoulos and Widmer[1] pointed out that the industrial sector of the U.S. economy operates at an overall second law efficiency of about 13%, leaving considerable room for improvement. These authors and co-workers also showed that approximately 25% of the projected 1985 energy usage in the manufacturing sector of the U.S. economy could be saved through conservation measures with capital costs on the order of $125 billion. By comparison, they estimated that to produce the same amount of energy from grass roots sources would require a total investment of about $175 billion. They quote an American Gas Association report[2] that indicates that, apart from the Middle East, the current capital investment required to provide new supplies of energy is very high. For

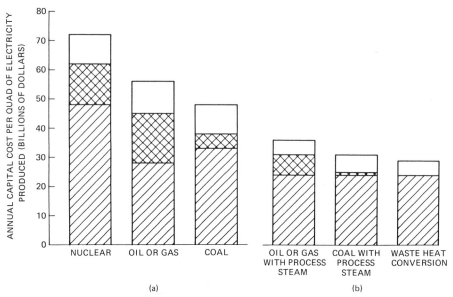

Fig. 10-1 Comparison of the capital costs of (a) central power station systems and (b) industrial cogeneration systems: □, transmission and distribution facilities; ⊠, fuel production facilities; ▧, generating plant. From Reference 3.

example, each new source of petroleum or natural gas equivalent to 456 kB/d costs between $5 and $10 billion in investment. A similar amount of "synthetic" gas or liquid fuel derived from coal would be even more costly, on the order of $15 billion. Although equivalent new coal supplies are still obtainable at a capital cost of about $2 billion, environmental restrictions at the mine and the user's site and the investment required for transportation equipment add significantly to this cost.

The capital cost for industrial energy is highest when the energy is provided in the form of electricity. In its 1977 annual report, Thermo Electron Corporation summarized its studies of the capital cost of central station power systems compared to industrial cogeneration systems for electricity production. The comparison is shown in Fig. 10-1.[3] The total investment for central station power systems using coal or oil is on the order of $50 to $60 billion/$10^{15}$ Btu · yr (one quad) of electricity. This includes transmission and distribution facilities, fuel production facilities, and the majority of the costs of the generating plant itself. Industrial cogeneration, on the other hand, requires only 50 to 65% of this investment, because the total investment is shared with onsite users of process heat that would otherwise need a separate supply. Obviously, the nation would be far better

Table 10-1

Examples of the Cost Effectiveness of Conservation Measures[a]

Conservation step	Energy saving (10^6 Btu/h)	1975 capital (thousands of dollars)	Annual quad fuel (billions of dollars)	Capital for comparable new supply[b]
Place recuperaters on the steel reheating furnaces (170 ton/h)	100	480	0.7	$10 \times$ Cons.
Optimize the crude product train (100,000 B/d)	23	1000	4.7	$1\frac{1}{2} \times$ Cons.
Add feed water heaters on the styrene plant furnace (400 ton/d)	4	17	0.6	$4-12 \times$ Cons.
Generate steam from the flue gas of the NH_3 reformer (1000 ton/d)	11	130	1.1	$2-6 \times$ Cons.

[a] From Reference 1.
[b] Cons. stands for the capital required for the conservation step.

off if industrial cogeneration investments were substituted for grass roots fuel production and central station power generation.

These global studies point out the economic potential of energy efficiency in the United States. Although other countries may be more productive in the use of energy in the industrial sector, appreciable potential exists for them as well. The practical difficulty with the implementation of global studies is that institutional barriers play an important role in society, but are ignored in the studies. Industrial managers function within an assigned area and work very hard to maximize the profit impact of their sector. They get no credit for savings achieved by others and have little time to spend in persuading other parts of their (or other) organizations to go into a cooperative venture in the energy field. In spite of the global potential for energy conservation ideas specific to their own area (see Table 10-1 for a few examples), the perceived rewards are not high enough to merit action. Table 10-1 shows that the total capital cost for new supplies often exceeds that for conservation measures. Still, much of the capital costs considered for new supplies would have to be spent outside the industrial plant gate. Therein lies part of the problem for the industrial energy conserver.

Typical of this situation was the case of a plant manager who proposed to install several million dollars of cogeneration equipment powered solely by waste heat. This investment would have replaced at least $5 million of investment in a central station power plant and would have provided a 22% return on investment at an electricity price of 2.6¢/kW·h. Despite the fact that this return was more than twice the company's average return on assets, top management rejected the proposal. The reasons given were two-fold:

(1)   the plant management was not faced with the $5 million investment for power generation facilities;

(2)   The management had set a 30% return on the investment hurdle rate for investments not related to the production of product.

By company practice the manager was forced to use the average cost of electricity rather than the replacement value in evaluating his project, and could not take credit for investment savings by others. As a result his project, however beneficial to the nation as a whole, was turned down.

Thus, the institutional nature of the industrial organization presents additional problems for the energy conservation engineer that are not faced by global policymakers. The thrust of this chapter will be to discuss some principles that *do apply* within the gates of a single industrial establishment or that might be achieved by the cooperation of a few industrial organizations located in the same area.

I. The Entire Plant Energy System is Pertinent

A. Capturing System Effects in Capital Analysis

As discussed in Chapter 6, many energy conservation steps cause thermodynamic interactions in the plant energy system. Each of these interactions has an impact on the capital cost as well as the fuel consumed in the energy system. In grass roots situations the size of boilers, steam piping, water treatment plants, fuel lines, scrubbers or precipitators, cooling water systems, etc., are directly affected by demand, at least up to the point at which company policy mandates an irreducible minimum spare capacity for inspections and the like. In systems with a large proportion of facilities already in place, the impact may not be as direct but can still be significant. Older, less efficient boilers may be shut down, additional spares not purchased or installed, surplus lines used for other purposes, excess heat exchange surface used to improve efficiency, excess utility production sold to neighbors, etc.

Table 10-2

Lost Availability Comparison of the Conventional (Fig. 6-4a) and Cogeneration
(Fig. 6-4b) Cycles

Parameter	Conventional cycle	Cogeneration cycle
Availability input (Btu)		
Fuel at 21,370 Btu/lb	29,362	21,648
Boiler feed water at 344°F and 125 psia	1,071	1,071
Total input	30,433	22,719
Useful output (Btu)		
Electricity	3,413	3,413
Steam	7,130	7,130
Total output	10,543	10,543
Total lost availability (Btu)	19,890	12,176
Second law efficiency	34.6%	46.4%

Thus, each energy efficiency improvement merits a reevaluation of the capital needs and utilization of the entire plant energy system, as well as those for the specific change.

In the process industries it is not uncommon to have related industrial operations grouped together at a single site. Refineries, basic petrochemical producers, and plastics manufacturers are a common grouping. Often these complexes generate at least a part of their own power requirement either as electricity or in direct mechanical drives utilizing high-pressure steam turbines and/or gas turbines with waste heat boilers. Utility systems in such plants are generally very conservatively designed, with more than adequate spare capacity to account for statutory inspections, power interruptions, and the like. In such situations cogeneration benefits similar to those outlined in the global studies can often be realized within the gates of the complex, if not within the gates of an individual member of the complex.

Figure 6-4a (p. 120) shows a hypothetical low-cost energy cycle supplying 1000 kW of electricity and 17,645,000 Btu/h of process heat in the form of 125-psia steam discussed earlier. This would represent a more or less typical approach in the refinery or petrochemical business, in which efforts were made to minimize investment because energy was inexpensive. Because high-pressure boilers have relatively high capital costs, the size of this system was generally minimized by operating a condensing turbine. The low-pressure steam was raised separately in a less expensive boiler and used for process heat. The alternative cogeneration approach is shown in Fig. 6-4b (p. 120). This was shown to be more efficient in the previous discussion (see Table 10-2). In this system approximately twice as much high-pressure

Table 10-3

Investment Comparison (1981) of the Conventional (Fig. 6-4a) and
Cogeneration (Fig. 6-4b) Cycles[a]

Component	Conventional cycle	Cogeneration
Pro rata high-pressure boiler[b] and turbogenerator system	2180	2530
Condenser and cooling water system	200	—
Pro rata low-pressure boiler[c]	520	—
Kettle reboiler	—	270
Total	2900	2800

[a] All values are in thousands of dollars.
[b] Based on a 200-kpph, field-erected, 1600-psig, 950°F boiler.
[c] Based on a 100-kpph, packaged, 150-psig, saturated boiler.

steam is required, because the power is generated in a turbine in which the exhaust steam is extracted at 200 psia rather than at the condensing pressure. The high-pressure boiler is the only one that is required, however, because the exhaust steam provides the process heat. Also note that the cooling water system required for the condenser in Fig. 6-4a is no longer necessary.

The grass roots capital cost for the entire cogeneration system is also slightly lower (see Table 10-3). However, if only the cost of the high-pressure boiler were considered, one would get a different answer. *Care must be taken to evaluate the capital cost of all the affected parts of the system on a consistent basis.* In developing this table, the author has assumed that pro rata costs of a larger boiler, turbogenerator, and cooling water system would be applicable in both cases. The required investment for the more efficient system is no worse than a standoff compared to the system that was thought to require low investment. Thus, the potential exists for developing grass roots cogeneration systems within the gates of a plant, or a complex of plants, at no increase in overall plant capital cost while achieving the appreciable energy savings outlined in the global energy studies.

To those intimidated by the thought of even the simple interaction of running a steam line across the fence to a neighbor's plant, it should be pointed out that many of your competitors may have been doing this kind of thing for some years. Often such arrangements can be made with the local utility, which is generally happy to supply steam to a customer on a shared benefits basis. Gas turbine drivers on pipelines or in industrial plants may also generate surplus steam from the exhaust or provide preheated (slightly O_2 depleted) air for other combustion operations at the same site. In many cases the service factors of integrated units can be decoupled by simple devices, making interaction possible without an intolerable number of constraints.

Environmental considerations often facilitate this kind of interaction as well. The shutdown of old, inefficient boilers may provide offsets against which new production facilities can be built. Similarly, current problems may be eliminated by shutting down equipment made superfluous by the integration. Although all benefits ultimately boil down to an economic contribution, looking outside of the energy system for credits is often very profitable.

B. Large Single Plants Can Benefit Independently

Many times those people least inclined to cooperate in cogeneration ventures come from your own company. With persistence, the corporate general interest will usually overcome. Simplified presentation of the overall benefits, uncluttered by who gets what share of the benefits according to current accounting practices and an audience with higher management are generally required to gain approval.

To put the hypothetical case into a practical perspective, let us consider the example of a major petrochemical plant addition to a refinery–chemical plant complex. The new plant was to be located about a mile from the existing facilities and had the potential to export a large amount of medium-pressure steam if its boilers were built with adequate capacity. Figure 10-2 summarizes the overall benefits for both the petroleum and chemical management that led to installation of the system. Corporate energy savings of several million dollars per year won top management support. A system for sharing the benefits based on booked investment was developed. Although occasional territorial disputes arise at turnaround times, all parties of the corporation are reasonably satisfied with the arrangement, as demonstrated by subsequent increases in steam export.

Large single plants with integrated steam power systems often have enough internal flexibility to reap combined capital and energy savings. An overall site plan is needed to accomplish this, as discussed in Chapter 6.

Consider the case of a grass roots ethylene plant to be built on a remote site. As demonstrated earlier, ethylene plants burn a great deal of fuel in the process furnaces and require a large amount of horsepower for process gas and refrigeration compressors. The base-case steam power system proposed by the contractor responsible for the process design of the plant is shown in Fig. 10-3. The waste heat boilers on the furnaces and the offsite boilers were planned to produce steam at 1500 psig that would be used to drive the two

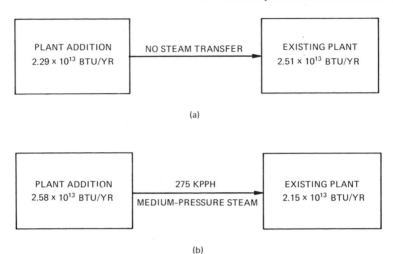

Fig. 10-2   Corporate impact of steam integration: (a) without integration; (b) with integration. The fuel savings would be 0.07×10^{13} Btu/yr; assuming \$3/MBtu, this corresponds to \$2,100,000/yr.

major plant compressors. Numerous turbines were operated with 600- and 200-psig steam, and process heat loads were integrated with the steam balance to maximize cogeneration opportunities. Still, approximately 155 ton/h of high-pressure steam flowed to the condensers on the turbine exhaust and about 1.4 billion Btu/h were fired in the furnaces and the boilers.

As discussed in Chapter 6, in such circumstances gas turbines often supply energy credits by reducing the amount of steam going to the condensers and the total amount of fuel fired at constant power generation. The owner's energy conservation engineers developed an alternative steam power system involving a gas turbine. This is shown in Fig. 10-4. The entire steam power system was reworked. Dual steam turbine drivers with different throttle and exhaust pressures were proposed for the propylene refrigeration compressor (PRC). A helper steam turbine was required for the gas turbine driving the process gas compressor (PGC), and the gas turbine exhaust was used as preheated air for the plant boilers, which were reduced to 600-psig. Because the boilers were much reduced in size, less steam was needed to drive pumps, fans, and other auxiliaries. Except for the deaerator, heat loads remained the same. The net result of the new balance was that the flow of steam to the condensers was reduced to 70 ton/h and the total fuel requirement to about 1.2 billion Btu/h. Fuel (available energy) savings of almost 200 million Btu/h were generated by this revised system.

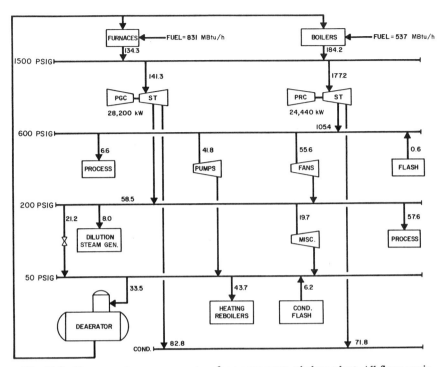

Fig. 10-3 Base-case steam power system for a grass roots ethylene plant. All flows are in tons per hour. The total amount of condensing steam is 154.6 ton/h and the total amount of fuel consumed is 1368 MBtu/h. PRC, propylene refrigeration compressor; PGC, process gas compressor; ST, steam turbine.

The key element in this example is that a small capital savings was realized as well. An investment summary is shown in Table 10-4. Including extra contingencies for the gas turbines and the boilers and almost $7 million worth of fuel system modifications because the plant was now in fuel surplus and needed to export, the net change in capital was about $1½ million of savings. By fully exploiting an understanding of the principles of energy efficiency and the interactions in its own steam power system, this plant was able to capture large energy efficiencies for zero incremental capital costs.

Again, considerable management debate was required to gain acceptance for this proposal. We reiterate this point to disabuse those who feel that a superior technical and economic solution will command management support spontaneously. It has been my experience that an appropriate sales effort is always required, even for sophisticated management. Do not despair, but rather, persist (intelligently)!

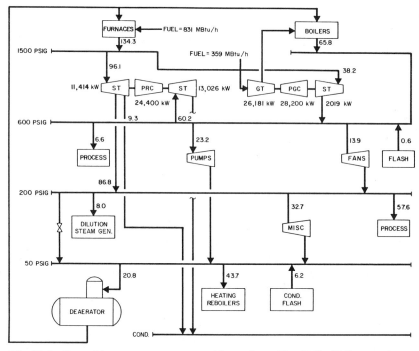

Fig. 10-4 Gas turbine steam power system (alternative to Fig. 10-3). All flows are in tons per hour. The total amount of condensing steam is 69.5 ton/h and the total amount of fuel consumed is 1190 MBtu/h. PRC, propylene refrigeration compressor; PGC, process gas compressor; ST, steam turbine; GT, gas turbine.

Table 10-4

Investment Summary for a Switch to the Gas
Turbine Driver of Fig. 10-6[a]

Component	Addition	Deletion
Process gas compressor		
Steam turbine	2,300	4,900
Gas turbine	8,350[b]	—
Refrigeration compressors		
Steam turbines	6,350	3,700
Steam system		
Waste heat boiler	11,350[b]	—
600-psig boilers	7,050[b]	—
1500-psig boilers	—	34,960
Fuel system modifications	6,600	
Total	42,000	43,500

[a] All values in thousands of dollars. Based on 1978 dollars.
[b] Includes a 5% extra contingency.

C. Cooling Water Systems Offer Opportunities

Cooling water systems have long been designed by "rules of thumb" based on practical experience. Sometimes the rules apply to all sites in a multiplant company. For a cooling tower system, they generally include a fixed temperature rise in the cooling water and an "optimum" approach to the design wet bulb temperature at the site. R. A. Crozier, Jr., proposed an alternate approach that promises savings in both the capital costs of the cooling water system and the energy required for pumping.[4] For the first time cooling water is viewed as very expensive and analyzed as one would a process system.

A typical cooling tower system is shown in Fig. 10-5. Heat is dissipated to the atmosphere by evaporating some of the return water. For a fixed tower outlet temperature, the higher the temperature of the water returning to the tower, the less water circulation is required, the smaller the required cooling tower, and the lower the air flow required in the tower. The latter is true because the water carrying capacity of the air increases appreciably with temperature. Obviously, we are trading lower investment in the cooling tower and cooling water circulation system for larger investment in the heat exchangers and potentially more corrosion and fouling problems.

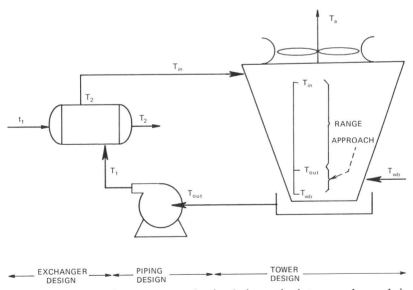

Fig. 10-5   Typical cooling tower system showing the interaction between exchanger design, piping design, and tower design. From Reference 4. Excerpted by special permission from *Chemical Engineering* (April 21). © 1980 McGraw-Hill, Inc., New York, N.Y. 10020.

Crozier empirically determined that if the log mean temperature difference (LMTD) of an exchanger is greater than 30°F, its area will be relatively insensitive to coolant temperature. In such cases savings in investment and energy in the cooling water system are indicated.

The tower outlet temperature is the most sensitive variable in the system, particularly in refrigerant or condensing turbine applications in which very low LMTDs are common. Two opposing constraints must be satisfied:

(1)   a closer approach to the wet bulb temperature of the air increases the cooling tower investment;

(2) a closer approach to the limiting process temperature increases the exchanger area.

On occasion it is more economic to provide separate cooling water systems for low-LMTD application rather than place stringent restrictions on the entire cooling water supply.

Crozier's system design will not be reproduced here. His system example will be discussed, however, to demonstrate two things:

(1)   the costs of cooling water systems are appreciable;

(2)   intelligent design, as opposed to "rules of thumb," can save both capital and operating costs.

A five-exchanger system totalling 90 MBtu/h was evaluated on two bases, Crozier's guidelines and a "rule of thumb." The "rule of thumb" case assumed a coolant temperature rise of 20°F unless a temperature cross resulted, in which case a 10°F approach to the process outlet was substituted. In the "guidelines" approach, the coolant outlet temperature was set equal to the process outlet temperature. For four of the five exchangers, the LMTD is > 30°F and interactions need not be considered on a practical basis. For the fifth exchanger, a real optimum could be determined by plotting capital and operating cost against coolant rise. However, from the practical standpoint, a "near-optimum" situation can be obtained by following the "guidelines" for LMTD > 30°F for all five exchangers. The results are shown in Tables 10-5 and 10-6. Because the coolant flow rate is reduced by half, more than $300,000 in capital costs and about $90,000/yr in operating costs are saved in the "guidelines" approach. A relatively small percentage of these operating savings will be offset by increased treating costs to control fouling in the system at higher temperatures.

Thus, it seems fair to conclude that antiquated design rules for cooling water systems need to be reevaluated in today's environment. Reducing the total cooling water flow rate would seem to be more important economically than minimizing heat exchanger area.

Table 10-5

Process and Investment Data for the Application of Design Guidelines to a Five-Exchanger Cooling Water System

Exchanger	Process temperature (°F) In	Process temperature (°F) Out	Rule-of-thumb design Coolant Outlet temperature (°F)	Flow rate (gal/min)	Exchanger investment (dollars)	Guidelines design LMTD (°F)	Coolant Outlet temperature (°F)	Flow rate (gal/min)	Exchanger investment (dollars)	Incremental differences Coolant flow rate (gal/min)	Exchanger investment (dollars)
1	212	151	108	1,800	35,000	62	150	600	39,000	−1200	4,000
2	212	113	103	2,400	45,000	52	113	1,600	52,000	−800	7,000
3	194	131	108	1,800	41,000	51	131	900	47,000	−900	6,000
4	158	113	103	2,400	50,000	33	113	1,600	63,000	−800	13,000
5	140	104	94	6,000	70,000	23	104	2,600	86,000	−3400	16,000
Total				14,400	241,000			7,300	287,000	−7100	46,000

Table 10-6

Cost Comparisons for the Application of Design Guidelines to a Five-Exchanger
Cooling Water System

Cost comparisons	Rule-of-thumb design	Guidelines design
Tower inlet temperature (°F)	100	115
Tower outlet temperature (°F)	88	90
Coolant flow rate (gal/min)	14,400	7,300
Tower, pump, and piping investment (dollars)	1,100,000	700,000
Water treatment investment (dollars)	230,000	220,000
Total coolant investment (dollars)	1,330,000	920,000
Incremental coolant investment (dollars)		−410,000
Incremental exchanger investment (dollars)		46,000
Total investment *saved* (dollars)		364,000
Coolant operating cost (dollars/yr)	210,000	110,000
Increased water treatment cost (dollars/yr)	0	10,000
Total coolant operating cost (dollars/yr)	210,000	120,000
Total operating cost saved (dollars/yr)		90,000

D. Impact of Process Changes

As discussed earlier, many changes to onsite energy use have an impact on
the site energy system. These interactions can often be reflected in the
investment required for the energy system.

Consider the demethanizer tower system shown in Fig. 9-11 (pp. 222)
from the point of view of identifying opportunities for saving energy. The
two new heat exchangers used to match the third feed temperature to the
tray temperature save 2200 hp in the refrigeration system. This corresponds
to a savings of about $400,000/yr in fuel. A rough estimate of the investment
was also $400,000, indicating a very attractive project on a retrofit basis.

At the process design stage, the investment in all of the utility systems
could also have been affected by this saving. Smaller refrigeration compressors could have been included; smaller drivers, smaller boilers, and, because
the last increment of power in the system must be supplied by condensing
turbines, smaller condensers and cooling water systems would also be
needed. One can argue that compressors and drivers tend to come in
increments, so that a 5% change may not affect the casing size and, therefore,
would have little impact on capital cost. This is certainly not true for large
high-pressure boilers, cooling water systems, and the like. Earlier we developed a cost of about $2500/kW for an industrial condensing power system
(see Table 10-3). If savings in the refrigeration compressors and system are

ignored and this figure is used to approximate the investment saving potential in the utility system for a 2200-hp (1650-kW) reduction in demand, we find a saving of $4,100,000!, as shown in the accompanying tabulation.

Improvement	Incremental investment (thousands of dollars)
Additional heat exchangers	400
1650-kW savings in steam system at $2500/kW	−4100
Net impact	−3700

You may argue with the figure of $2500/kW, but at any reasonable value for incremental investment in the energy system, this energy efficiency improvement has a negative impact on overall plant investment. Thus, early analysis of the demethanizer tower (i.e., when utility system design was still flexible) would have made possible a significantly larger impact on plant economics than the retrofit case.

E. Be Alert for Organizational Cracks

Why do opportunities like the demethanizer situation go unnoticed in the process design stage? The obvious reason is that the process designer did not have enough experience or lacked the correct data to identify it. This is possible, and it can be headed off in many cases by insisting on doing the thermodynamic analysis during the process design phase or immediately after design completion as part of the review process. This adds to the engineering cost and may be fought on the basis of scheduling as well, so do not expect immediate acceptance of the idea.

Another explanation, however, is that the process designer saw the opportunity but decided it was not attractive. Obviously, the designer did not have the data presented earlier if that conclusion was reached. In my experience there are three potential reasons for faulty evaluation:

(1)   schedule demands;
(2)   poor organization of the design team;
(3)   acceptance of suggestions by operating personnel without economic evaluation.

Design schedules are funny things. Projects, particularly major projects, are dreamed about, thought about, and then seriously planned for years. Once that is done, the design has to be completed in 6 months, and no changes are tolerated once mechanical design and purchasing have begun (unless they come from operating personnel reviews). People (around management) are always amazed when hindsight or cost-cutting programs identify opportunities not included in the process design of a project. At least part of the answer is inadequate time allowed for experienced designers to step back and analyze their efforts.

There are many ways to combat this problem. One can do a lot of process design in parallel with project planning and basis setting. One can program in a period of process design review aimed at optimizing the whole once the design is complete. One can set up systematic reviews during the process design effort. The list is endless and should be a function of each company's project execution system.

Design teams are usually organized by skill groups. Separation process engineers, furnace designers, and steam power experts rarely function in a cohesive unit. A common way to promote optimization in the design is to establish incremental values for the utility streams to be used in optimization studies. Thus, when a fractionation engineer wants to trade-off trays versus reflux, a value for steam supplied by the steam power experts is used. Similarly, the furnace designer evaluates steam produced in the convection section at these prices.

The problem with this approach, as necessary as it is, is that the utility values supplied by the experts are often very preliminary (after all, the steam power people do not really know what the system will look like, because the loads are not well defined yet) and do not adequately reflect the investment impact on the whole utility system. Energy, water, and chemical costs are relatively easy to project, but without an approximate plant layout, a reasonable steam power balance, rational driver selection studies, and full input from the cost estimators, it is very difficult to provide a full description of the impact of utility investment early in the project.

The best efforts are made when the cost engineers set up separate guidelines for incremental investment in the total system early in the project, including carefully defined limits of applicability. Then an onsite process designer can better identify the net overall plant investment impact of any change that is contemplated. In addition, the cost engineer will have an ownership interest in providing numbers that will stand up at the end of the design when detailed project estimates are made.

Operating reviews invariably do two things: improve project startups, safety, and operability and add cost. Rarely are the reviewers asked to justify the changes suggested (demanded?). Requirements for backup utility systems, bypasses, maintenance access, etc., often put restrictions on the designers because of the constraints imposed on potential utility system credits. There needs to be as serious an economic analysis of operating department suggestions on the overall plant cost as that carried out for process design improvements.

This discussion is not meant as the tirade of a frustrated process designer or energy conservation engineer. Anyone who has functioned in these areas for a while is aware of the constraints imposed by others on the purity of one's function. Planning ahead and active agitation early in the project to establish the most favorable conditions for the design seem to be the only practical ways to provide the data and time to do an optimum job.

II. Investment Optimization

A. Accepting Practical Constraints

Energy conservation ideas are often based on steady-state operations and predicated on achieving a high service factor to match the base operation. In some cases this may not be optimum or even justifiable. A partial solution or a part-time solution may often be had at much lower investment and much higher profitability. This is partly a follow-up on the theme of Chapter 5, "get the most you can out of what you have."

To illustrate this point let us consider a practical situation. An overall site energy survey suggested converting a number of reboilers on a distillation tower from 110-psig steam to 40-psig steam. Obviously, using a lower available energy source to supply process needs is thermodynamically sound. One tower was in a fouling service and had a spare reboiler so that cleaning of the heat exchanger could be accomplished without lost production. The current arrangement, in which two reboilers operate in parallel with the third held as spare is shown in Fig. 10-6a. Fig. 10-6b shows a simple extrapolation scheme in which additional surface is provided to maintain the same spare surface ratio in the 40-psig case. In the 110-psig case, each bundle is removed for cleaning once the steam side temperature reaches 312°F (65 psig), because operation beyond that point causes tube plugging. The average bundle run length over the last few years has been 6 months. Two to three days are required to clean one bundle.

The present performance of the reboilers is shown by the dashed curve in Fig. 10-7. The fouling rates increase with time, reaching a fouling factor of 0.005 in 4 months and end of run (0.01) in 6 months when the steam side temperature is 312°F. This is consistent with previous general experience with polymer-type foulants indicating that growth is slow on clean metal surfaces and then accelerates as the polymer covers the surface.

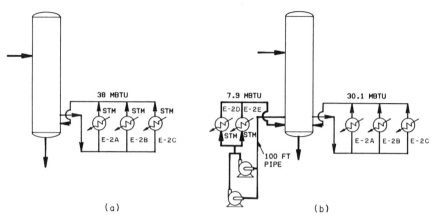

Fig. 10-6 Reboiler conversion for investment optimization: (a) original plan using 110-psig steam; (b) first revised plan for using 40-psig steam. Exchangers E-2A,B,C, 5600 ft²; exchangers E-2D,E, 7000 ft².

The original plan (Fig. 10-6) was to add reboilers to keep the same ratio of operating to spare surface with 40-psig steam and the same run length. This meant adding two more reboilers in parallel with the first two at a projected cost of $1 million. The return on the project was well in excess of management hurdle rates.

Further analysis led to a much more economic solution. The total surface already available (including spare) was adequate for operation with 40-psig steam if it were all placed in service simultaneously. For an operation with all three reboilers in service on 40-psig steam, the maximum steam side temperature would be 284°F. Calculations show that a fouling factor of only 0.005 could be tolerated to achieve the required duty. Without taking any credits for reduced foulant formation, the operation would proceed as shown by the solid curve in Fig. 10-7. The bundles would reach a 0.005 fouling factor in 4 months. Then the reboiling would return to 110-psig steam for 6–9 days while each bundle was cleaned. With this procedure, there would be a loss of 40-psig steam credits for up to 27 days a year. On the basis of a $2.24/klb difference in steam costs between the two pressure levels this would amount to $35,000 lost on a potential $410,000/yr, but capital costs would be almost zero.

There are indications that the exchangers using 40-psig steam would have run lengths longer than 4 months. Steam side temperatures for the three-bundle case are 7°F lower than those for the two-bundle case at the time that the fouling factor builds up to 0.005. This favors a lower polymerization rate. Existing correlations (although not exactly applicable) would predict that run length would increase by a factor of 1.3 or more. Also, heat density

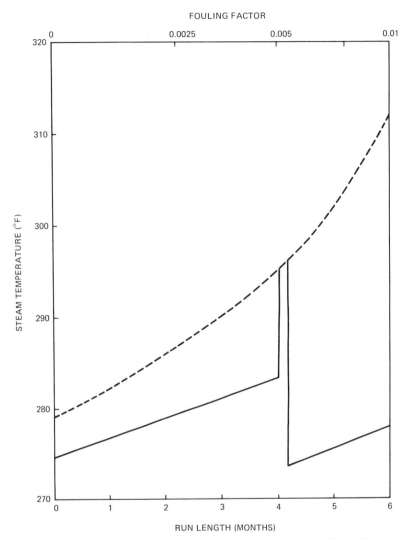

Fig. 10-7 Present fouling performance of the reboiler arrangement of Fig. 10-8b (---, two bundles) and fouling performance for a more economic solution of the reboiler arrangement of Fig. 10-8 using 40-psig operation on existing exchangers (——, three bundles).

is reduced in the 40-psig steam case. This directionally reduces fouling because it guarantees a wetter wall condition.

A no-risk test is available to validate the predicted operation before the 40-psig steam system is placed in operation. The spare bundle could be placed in service and steam side temperature monitored. When the temperature reaches 284°F, equivalent to operation with 40-psig steam, one bundle could be taken out for cleaning and returned to spare status. The other two bundles would have two more months of service before reaching the 312°F steam temperature limitation. Actual operation should provide even better results than this simulated test, because temperatures would be lower. Once the 40-psig steam system is placed in operation, the maximum steam temperature that the outlet end of the reboiler tubes would be exposed to is 287°F. This should make for a significant reduction in fouling at this point.

With a more sophisticated control system and more thorough analysis, the operating department accepted this new approach to operating the tower. The net capital cost was ∼10% of the original idea but >90% of the savings potential was achieved. The operating people accepted a more complex procedure in place of more of the same because of its economic potential.

Testing all investments for partial solutions may yield similar economic improvements. A particularly fertile option to consider is piggybacking on maintenance programs to get small improvements in performance for very little increase in the already budgeted maintenance expenditure. For example, if a reactor is to be rehabilitated, an improvement in heat recovery, conversion, or selectivity might be accomplished by small changes in design, residence time, or catalyst properties at very little increase in cost.

B. Spend for Performance, not Structure

In a number of cases there are choices between improved performance (which may represent departures from normal practices) and enhanced reliability or convenience. Sometimes the cost of backup equipment or structural steel will add up to the same total as a scheme offering improved performance.

Consider the waste heat boiler arrangement shown in Fig. 10-8. A waste heat boiler was to be added to five furnaces to reduce stack temperatures by generating 130-psig steam saturated at 355°F. To provide the required pressure drop through the boiler by natural draft, a 245-ft high stack was required (the maximum permissable), and stack temperatures had to be maintained at 450°F. This provided a large temperature difference in the waste heat boiler and limited the heat recovery potential of the system.

As indicated earlier, such constraints have to be tested, especially when large expenditures are contemplated because of them. An induced-draft fan was proposed to eliminate the limitations of the natural-draft system. This is shown in Fig. 10-9. The extra pressure drop permitted both a closer ap-

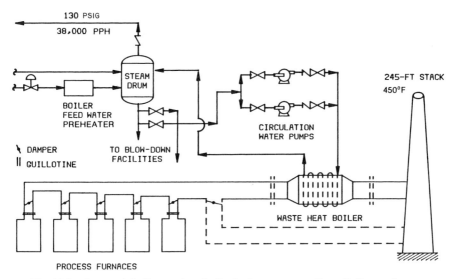

Fig. 10-8 A natural-draft waste heat boiler for heat recovery. From Reference 5.

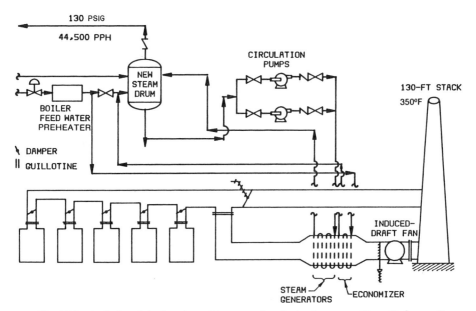

Fig. 10-9 An induced-draft system with economizer for heat recovery. From Reference 5.

Table 10-7

Multifurnace Waste Heat Boiler Study for a Natural Draft (Fig. 10-11) and
an Induced-Draft (Fig. 10-12) System[a]

Parameter	Natural-draft system	Induced-draft system
Stack height (ft)	245	130
Duct velocity (ft/sec)	25	25
Draft	Natural	Induced
Fan horsepower	—	165
Boiler pressure (in. H_2O)	0.5	7.5
Steam capacity (PPH)	38,000	44,500
Economizer?	No	Yes
Feed water temperature (°F)	267	303
Stack temperature (°F)	450	350

[a] From Reference 5.

Table 10-8

Economics Summary of a Natural-Draft (Fig. 10-11) and Induced-Draft
(Fig. 10-12) System[a]

Cost or savings	Natural draft (38,000 PPH)	Induced draft (44,500 PPH)
Investment		
Stack	900,000	600,000
Ducts	500,000	500,000
Structures	305,000	290,000
Waste heat boiler	375,000	400,000
Steam drum	90,000	90,000
Fan	0	80,000
Economizer	0	115,000
Engineering, contingency, and other	1,480,000	1,585,000
Total	3,650,000	3,660,000
Unit cost per MBtu/h	95,000	80,000
Annual operating cost savings		
Steam at $3/1000 lb	940,000	1,100,000
Boiler feed water	(80,000)[b]	(100,000)
Power	(10,000)	(25,000)
Maintenance	(60,000)	(75,000)
Total	790,000	900,000
Payback period (yr)	4.6	4

[a] All monetary values are in dollars. From Reference 5.
[b] Savings in parentheses indicate extra costs.

proach to the process temperature and the addition of an economizer to enhance recovery further. A comparison of the performance of the two cases is shown in Table 10-7.

Rough capital estimates (1977) are shown in Table 10-8 along with a summary of economics. The fan scheme trades investment in stack (a constraint) for investment in improved efficiency, which will be more valuable as energy costs increase. The difference is that in a power failure benefits might be lost for a short time. Neither project is an award winner at $3/1000 lb steam, but the investment per million Btu per hour saved is significantly better for the fan project. Project economics might be further improved by optimizing the fan project within the new constraints.

C. Retrofits and Energy System Capital

As discussed earlier, retrofit energy conservation projects face steeper challenges than grass roots improvements. This is not only because they do not share in the economies of scale of the base project, but also because additional constraints are placed on obtaining credits for investment savings in the energy system.

Obviously, savings in sunk investment are only of value when alternate uses for that investment are available. If the steam system must be expanded to meet other additional loads, then liberated boiler and cooling water capacity would be of value. Alternatively, older capacity might be nearing retirement, and investment credit would be gained by eliminating the need to replace or refurbish obsolete capacity to maintain system reliability. For the case in which there is no current use for the liberated utility system capacity, the economics of the retrofit could not be credited with the savings generated by reduced loads, unless the plant accounting method allocated utility investments to the users on the basis of use. Under such circumstances, some share of these savings would accrue to the improved operation, depending on what other utility system users existed on the site. Thus, for retrofit investments other circumstances will dictate what credits, if any, truly apply for sunk investment.

Consider the scheme for using the heat once rejected in the condenser of a xylene tower to reboil benzene and toluene towers, whch is shown in Fig. 10-10. The original steam reboilers on the two product towers were replaced with larger heat exchangers suitable for condensing the xylene tower overhead. The overhead condenser of the xylene tower was discarded, and the tower operating pressure raised from 20 to 40 psig to provide appropriate temperature differences for the reboilers on the toluene tower. This slightly raised the heat duty (and the fuel requirement) on the xylene tower. The new

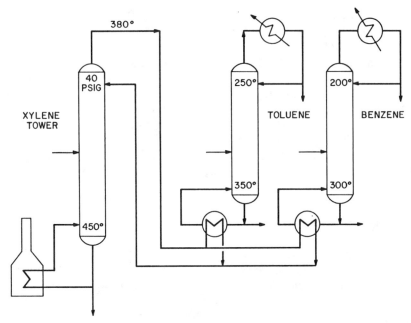

Fig. 10-10 Revised tower condenser and reboiler arrangement using the overhead vapors from the xylene tower to reboil two other towers.

utility consumptions and energy savings are shown in Table 10-9. Note that a small amount of additional pump horsepower was required to deal with the increased reflux flow and head on the xylene tower, but that overall savings of about $1.8 million/yr were achieved.

Table 10-9

Utility Consumption for Thermal Coupling versus Conventional Distillation

Utilities required	Conventional distillation (20 psig)	Thermal coupling (40 psig)	Operating savings ($10³/yr)
Fuel	126×10^6 Btu/h	140×10^6 Btu/h	(270)[a,b]
150-psig steam	97×10^3 pph	—	2500
Electricity	Base	72 kW	(25)[c]
Condensate	—	—	(410)
Total	—	—	1795

[a] Assuming a fuel cost of $2.3/MBtu.
[b] Savings in parentheses indicate extra costs.
[c] Assuming an electricity cost of $0.039/kW.

Table 10-10

Overall Site Investment Requirements for Thermal Coupling versus
Conventional Distillation[a]

Investment	Conventional distillation	Thermal coupling
Towers	Base	Base
Reboiler (incremental)	Base	1390[b]
Pumps and motors (incremental)	Base	600
Piping (incremental)	Base	310
Electrical supply (incremental)	Base	50
Cooling water system (incremental)	(900)	Base
Boiler systems (incremental)	(2600)	Base
Total	(3500)	2350

[a] All values in thousands of 1979 dollars.
[b] Includes investment for relocating some existing heat integration equipment.

Table 10-10 lists the incremental overall site investment requirements for the two cases. About a $2.4 million investment would be required for the extra heat exchangers and piping in the cascade system. Thus, the project is very attractive on an onsite basis alone. The liberated capacity in the cooling water system and in the boiler systems, results from the fact that steam demand is reduced by about 100 kpph and the xylene condenser is eliminated from the cooling water circuit, are listed as investment credits. These were calculated on a pro rata basis from larger system investments. If these credits can be taken, the savings in the utility system more than offset the cost of the onsite retrofit investment.

Again, the investment in the utility system far outweighs the investment in onsite facilities. Any part of this credit that can legitimately be claimed for the retrofit project will greatly enhance its economics. Time spent evaluating this aspect could be very profitable.

Some plants assign credits (or debits) for utility investment directly to projects on the basis of overall value to the site. So-called "associated facilities investments" or "pro rata boiler corrections" to an onsite investment proposal are the usual approach. This is possible at larger integrated sites where central steam power systems exist. As the sites grow, economic units of steam and power are added and the cost prorated among the users. Charges for unused capacity go into the overall cost of steam and power. Because these charges are both plus and minus, they present a mechanism for crediting retrofit projects with liberated utility investment.

Unfortunately, when sites are not growing, i.e., there are no likely customers for liberated capacity, this system breaks down. A serious utility marketing effort among industrial or commercial neighbors may revitalize it, if the plant management is willing to get into the business.

III. Defining the Limits of Current
Technology

A. Susceptibility to Capital

Some inefficiencies in modern processes are subject to being greatly reduced by spending more capital. For example, a pressure drop through piping can be reduced by increasing pipe size until the change in fluid pressure becomes negligible. As one progresses in this direction, a point is reached at which the total capital cost of the plant (including the cost of utility systems, of course) begins to go up. The same is true for heat exchangers. Both pressure drop and temperature driving force can be reduced at the expense of surface area until irreversibilities approach zero. Up to the point of minimum reflux, more trays in a distillation tower will decrease reflux rates and utility investment. Again, sooner or later the total investment in the plant starts to increase as the process approaches reversibility in whole or in part.

Other parts of processes cannot be made more efficient by additional investment. These parts have inherent amounts of lost work (lost availability, exergy) or entropy production because of the current state of technology. For example, the lost work associated with a heat of reaction (such as combustion) cannot be recovered because appropriate concentration cells are not available. On a smaller scale, if propane and propylene are to be separated by distillation, an inherent amount of lost work associated with minimum reflux requirements and a minimum pressure difference across the trays (necessary to maintain mixing and levels) must be tolerated, even if minimum reflux and minimum temperature differences in all heat exchangers are used.

Thus, we can separate most process irreversibilities into two categories: "capital sensitive" and "inherent." Clearly, some judgment must be made about the limits of current technology as assignments are made to each category.

B. The Infinite Capital Process

When the lost available energy of the entire process has been segregated in this manner, all of the irreversibilities sensitive to capital can be eliminated (or at least minimized) by increasing the size of equipment up to some thermodynamic constraint. The latter might be 1% over minimum reflux, 0.5°F LMTD, heat and material balance requirements, etc. This would give a plant that is uneconomic because capital is wasted on areas in which only a small percentage improvement in energy use is achieved. By the same token, the minimum energy consumption for existing technology is defined.

Use of this approach ensures that the optimum design is not overlooked. The existence of the cases described earlier in this chapter, in which both capital and energy cost can be saved without technological breakthroughs, is ample testimony that the usual case study approach to optimization sometimes fails.

A return to the discussion of Gyftopoulos and Benedict's air separation plant,[6] first mentioned in Chapter 9, is useful here. The original practical process and thermodynamic analysis were shown in Figs. 9-13 and 9-14 and Table 9-13. The authors then revised the process to minimize irreversibility when it could be eliminated by increased capital. In the idealized process, they assumed that 100% efficient machinery could be bought and sized the main heat exchanger so that its pressure drop and LMTD were minimum consistent with the constraints imposed by the heat balance and flow through the system. The losses in the distillation tower were essentially unchanged, however. The revised process was given in Fig. 9-14, and the new thermodynamic analysis was listed under the heading "ideal process" in Table 9-13.

The "ideal process" requires about one-fourth the actual work of the practical process, even though the ideal work is only slightly reduced. Work lost in the heat exchangers and machinery were the main areas of reduction. Let us focus on the main exchanger, as the authors did, rather than be distracted by the obvious impracticality of obtaining machinery capable of 100% efficiency.

As pointed out earlier, the authors showed that even with 1949 costs, the practical process was designed for too high a pressure drop (see Fig. 9-15). The equations developed by Gyftopoulos and Benedict for the process exchanger size, the power requirement, and the associated entropy change were used to explore the current onsite and power plant investments associated with reducing the entropy change of the process, i.e., with improving efficiency. Based on simplified investment curves, the average costs of equipment in 1981 dollars were calculated to be $5/ft² for heat exchange surface, $1300/kW for compressors and drivers, and $2500/kW for power generation.[7]

The investment costs at three exchanger temperature differences were calculated as a function of pressure drop (i.e., power requirement). Figure 10-11 gives the results of these calculations. The lower set of curves show that the investment for the heat exchanger and compressor alone generally decreases as the power requirement of the system increases. The three separate curves represent three different heat exchanger temperature differ-

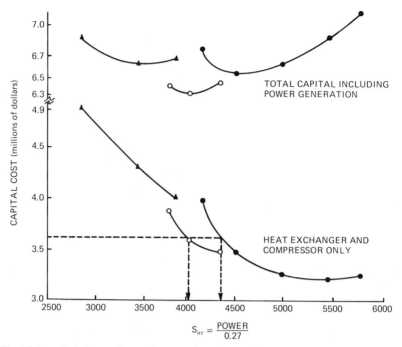

Fig. 10-11 Capital costs for an air separation plant at three exchanger temperature differences: ▲, 3°F; ○, 4°F; ●, 5°F.

ences. This is in line with the conventional expectation, *but* a 4°F temperature difference design has a 10% lower power requirement than either of the other choices at *the same investment level.* We can conclude that even when considering onsite investment alone, there is some potential to save both energy and capital.

In the upper set of curves the cost of onsite power generation for the compressor is added to the investment for the compressor and the heat exchanger to include energy supply effects. For any temperature difference selected for the heat exchanger, there is a rather pronounced minimum total investment point. This point corresponds to between 1.8 and 2.3 times the entropy change of the idealized process. Please note that the practical process operates at about 7 times this level of irreversible entropy change.

In addition, there is an optimum temperature difference for this exchanger that has a lower investment *and* better energy efficiency than either of the other two. Thus, the total investment for a 4°F driving force is lower than either the 3°F or 5°F curves. Also of interest is the fact that for the same investment, one can design an air separation plant to operate with a 3°F temperature difference and use about 75% of the power required with a 5° temperature difference. These results rest on the assumption that there are no significant changes in the investment of other components of the system as the configuration of the heat exchanger and compressor are changed within a relatively narrow range.

The unique contributions of Gyftopoulos and Benedict were to reduce a practical process to a simple energy model and to define the energy consumption of the process when not restricted by capital. This is much different than the classic "reversible process" of thermodynamics, and it provides a basis for assessing how much energy efficiency can be improved with conventional technology before capital costs skyrocket. From this simple analysis, energy consumption can be reduced from practical design levels to about twice that defined by the unrestrained capital case, before overall plant investment must increase to improve efficiency further. Because the practical design operates at about seven times the energy consumption of the idealized process, it seems that there is much potential to be explored before reaching this limit.

Qualitatively, these observations should also apply to other processes. However, except for several unrelated additional examples, such studies remain to be done. The approach of separating "capital-sensitive" and "inherent" irreversibilities and minimizing the former is recommended.

IV. Fundamental Process Improvements

A. Major Process Changes

Focusing research on a major "inherent" irreversibility can have a large impact on both energy and capital efficiency. These are the "new S curves" of those who create major process developments.[8] The S-curve concept stems from considerable analysis of process evolutions and breakthroughs of economic significance. The name comes from the typical shape of a benefit (presumably profit) versus development effort curve plotted over the life of a technology. A typical example is shown in Fig. 10-12. In the early phases of a new technology a great deal of effort is required to develop the necessary data base and forge a commercial niche. Once this is done, a very profitable phase is entered during which evolutionary developments make possible steep benefit curve changes for small incremental efforts. Ultimately, the other bend of the S curve is reached as the technology matures and is almost completely understood. Small improvements are more difficult to achieve

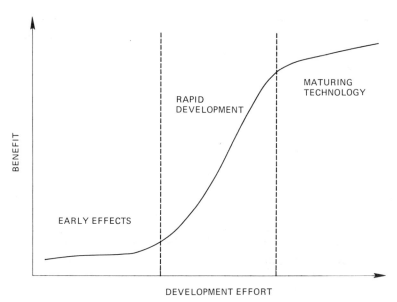

Fig. 10-12   A typical technology benefit curve.

and more research is devoted to finding a new technology to meet the same or expanded business need.

In Chapter 9 we discussed the potential of a concentration cell to derive work directly from the combustion step of a Synthane reactor. Others are studying the use of supercritical steam as an oxidation source. Although these particular breakthroughs may require some years to achieve, other significant improvements have been commercialized since 1973.

A major example is the Unipol process for a new polyethylene with improved properties, linear low-density polyethylene (LLDPE). The conventional process for low-density polyethylene (LDPE) involved a reactor pressure of about 50,000 psi. Reactions were carried out in 5000-ft long, 1-in. i.d. double pipe reactors with no energy recovery from either the heat of reaction or the pressure letdown. These conditions were required to obtain reasonable yields and economic reactor costs with the current catalyst and initiator systems. A considerable amount of evolutionary development occurred in the method of feed injection and reactor cooling to make small improvements in yields and capacity within the framework of known reactor technology. Soon the process entered the upper branch of the S curve.

Then came a new catalyst (really several). The reaction could now be carried out in a large fluid-bed reactor at a pressure of only a few hundred pounds per square inch. The first editions of this new technology needed 90% of the power, 10% of the steam, and only 85% of the capital of a commercial-sized plant relative to the older technology.

Large potential for future development also exists. The technology is truly entering the steep part of the new S curve. Currently pellets are still required to meet customer's needs, but powder technology is being developed that would reduce production costs significantly. The process itself contains large recycle streams, which might be made smaller by higher catalyst efficiency, etc.

Similar breakthroughs occurred when vinyl chloride monomer was produced from ethylene rather than acetylene, with the hemihydrate process for phosphoric acid, with the use of membranes to separate H_2 from gas mixtures, and in the substitution of plastic for much of the metal in automobiles in an effort to improve gasoline mileage.

Fundamental analysis of existing processes can do a lot to focus research on the major sources of lost work in a process. The ideal work needed to produce polyethylene from ethylene has not changed too drastically, but the efficiency of the process has, and markedly! Knowing where to attack improves the chance of success, even though a creative leap is still required and must be generated by the intuitive efforts of the researchers.

B. Phosphoric Acid — A New S Curve

Less dramatic process improvements can also have a significant impact on energy use. Conventional wet-process phosphoric acid plants digest phosphate rock in dilute acid and precipitate gypsum [calcium sulfate dihydrate $(CaSO_4 \cdot 2H_2O)$]. After filtering the solid from the acid, the latter is usually concentrated by evaporation to 50–55% purity for fertilizer manufacturing.

Nissan Chemical of Japan felt that much energy was wasted by adding dilution water in the digestion stage only to have to evaporate it later to produce salable phosphoric acid (good second law thinking!). Through study of the phase diagrams, they determined the conditions under which phosphoric rock could be digested to form the hemihydrate $CaSO_4 \cdot \frac{1}{2}H_2O$. This involved higher temperatures (95°C vs. 60–70°C) and 50% acid strength. By learning to filter the hemihydrate from the strong acid, a salable acid product could be produced without the need for evaporation. The hemihydrate was then recrystallized as gypsum, using recycled filtrates to wash the hemihydrate filter and provide water of hydration.

Heyward-Robinson Company, Inc., an engineering contractor based in New York City, has prepared some economic comparisons of the Nissan hemihydrate process with a "typical" dehydration process at a capacity of 1100 short tons per day. With purchased H_2SO_4, power requirements were about the same, but steam consumption was reduced from 0.7 to 0.1 ton steam/ton acid, or about 85%.

Onsite investments for the two grass roots plants also showed an advantage for the hemihydrate process of about $5 million (in 1980) out of $36 million. When the cost of additional steam generation was added, the saving increased to $10 million.

For plants in which H_2SO_4 is generated onsite, the waste heat boilers in the H_2SO_4 plant are normally used to supply the steam for phosphoric acid concentration. With the hemihydrate process, this steam is available for sale.

The point of this example is that some processes may offer the potential for significant improvements without laboratory breakthroughs. Focusing process development activities on a fundamental process inefficiency pinpointed by thermodynamic analysis may well offer a new S curve through engineering improvements only.

Summary

Fundamental studies indicate that the potential exists to reduce overall capital investment as energy efficiency is improved from present levels. There are also practical studies that show this is also feasible (up to some point) within the confines of integrated chemical complexes. Practicing conservation is often a lower investment case than obtaining new supplies. A key element is to be sure that the investment impact of onsite energy efficiency improvements on the plant utility system is explicity evaluated. Forming cogeneration ventures with industrial or commercial neighbors may also enable practical implementation of these opportunities. Optimization of the cooling water system should not be ignored.

The potential for joint energy and investment savings inside the plant gates is greatest for new grass roots designs. Here the greatest flexibility to optimize the whole site exists, and savings in utility investments can most easily be realized. Similar potential exists when utility system expansion is required for other purposes. In both cases cracks in the design and project organizations must be carefully monitored to avoid missing opportunities through imperfect communications.

In retrofit situations the value of energy system investment credits will depend on the external demand for liberated capacity. Partial solutions may be optimum in many cases.

Many other specific studies in industrial energy conservation programs have shown the potential for energy savings at zero or negative capital cost. Examples are presented in this chapter to demonstrate specific principles and to encourage energy conservers, not to provide a complete checklist of possible places to look for such projects.

There is a limit to the extent that process efficiency can be improved at lower total plant investment. Preliminary indications drawn from the air separation plant study are that this limit is probably far from the design point used in practical processes. More work must be done in this area to quantify the operative relationships. The approach suggested is to separate capital-sensitive and inherent process irreversibilities and minimize the former by enlarging equipment. This should result in a plant with nonoptimum capital investment but much reduced energy requirements. Investment can then be backed out until an optimum is reached.

Focusing process development studies on major identified irreversibilities holds the potential for major economic breakthroughs. Process step-outs, such as the new LLDPE processes, demonstrate some of the practical incentives for success in this area.

The intuitive nature of much of this discussion leads engineers to look for a more systematic method. Two attempts at this will be described in the next chapter.

Notes to Chapter 10

1. E. Gyftopoulos, and W. Widmer, Energy conservation. American Chemical Society Symposium Series 122, "Thermodynamics: Second Law Analysis," Washington, D.C., 1980.

2. A Comparison of Capital Investment Requirements for Alternative Domestic Energy Supplies. American Gas Association Report Planning and Analysis Group, Chicago, Illinois, May 1978.

3. Thermo Electron Corporation 1977 Annual Report, Waltham, Massachusetts.

4. R. A. Crozier, Designing a Near Optimum Cooling Water System. *Chemical Engineering* **87,** 118 (1980).

5. W. F. Kenney, "Waste Heat Recovery." MIT Process Series, MIT Press, Cambridge, Massachusetts, 1982.

6. E. Gyftopoulos, and M. Benedict, Air Separations Process. American Chemical Society Symposium Series 122, "Thermodynamics: Second Law Analysis," Washington, D.C., 1980.

7. Assuming an industrial condensing steam turbine generator with a condenser and a pro rata cooling water system.

8. These are the traditional Gompertz growth curves when growth is slow initially, then accelerates to an inflection point, after which it declines. Technology step-outs produce new curves displaced in the direction of profitability from the older ones.

11

Systematic Design Methods*

Introduction

In the previous chapters we have discussed how the use of data from various levels of process analysis can lead to the generation of ideas for process improvement. These ideas may be of maximum benefit only in grass roots process design, but in many cases they are suitable for consideration in the retrofitting of existing facilities. In all cases so far we depended on an experienced engineer to generate creative ideas based on intuition.

Various approaches have been under development for some time that are aimed at providing improved guidance to the engineer in finding an optimum process design. These approaches have attempted to integrate energy efficiency considerations with process constraints and economic realities so that no promising solution would be overlooked in the designer's effort. In many cases investigators have found opportunities to reduce capital costs as well as to increase efficiency, the potential for which was discussed in Chapter 10. These approaches have ranged from simple rules of thumb to complex mathematical representations of available energy changes and costs across the entire plant complex.

This chapter will discuss two of these approaches. The first is often dubbed "process synthesis" and frequently deals with optimum heat exchange networks. The second is usually referred to as "thermoeconomics," and it can involve very complex mathematical relationships.

* This chapter would not have been complete without the benefit of informal communications between M. Tribus of the Massachusetts Institute of Technology and J. Robertson of Exxon Research and Engineering.

I. Process Synthesis

Various investigators have attacked more or less standard heat exchange network problems by various systems, seeking to prove that they had found the optimum solution for the array. These problems have been labeled by shorthand for the various numbers of hot and cold streams. In this author's opinion, a great deal of a rather academic debate has gone on trying to prove that the best solution selected by a given investigator for a certain problem is indeed the *optimum* solution. Because very large numbers of solutions are possible for some of these problems, proving an optimum has turned into a very formidable task. The key issue for the practical process designer, however, is whether his solution approaches the optimum as closely as his time constraints and economic considerations allow. Thus, it is important to know approximately the best you can do in a given heat exchange situation.

The incentives to get close to the optimum arrangement have been demonstrated by Boland and Linnhoff[1] as well as many other investigators. They present an example for the front end of a specialty chemical process in the petrochemical industry. Figure 11-1a shows the traditional arrangement for the heat exchangers in this system. Six heat exchange operations are required, 1722 units of heat are supplied by utilities, and 654 units of cooling water are needed. In Fig. 11-1b a modified system (derived from the authors' systematic method) is shown that requires only four heat exchange operations and reduces the heat input to 1068 units. Fewer heat exchangers generally means lower capital cost, even if more surface is included in each unit. The new design not only reduces energy consumption by 40% but eliminates cooling requirements. Whether the system shown in Fig. 11-1b represents the optimum for the problem in question is moot. It certainly represents a significant improvement over the initial design proposed, and in the time frame available to the designer it may indeed be the practical "best solution."

A. Rules of Thumb versus Mathematical Search

Two general approaches to heat exchanger system optimization have been studied. The first involves rules of thumb that give step-by-step guidance to achieving a good answer. The second involves a listing of all possible solutions and a mathematical search through all of these possibilities to arrive at the optimum configuration. Because it is not infrequent that the number of possibilities approach 10^{18}, you can see that mathematical search is a very formidable task.

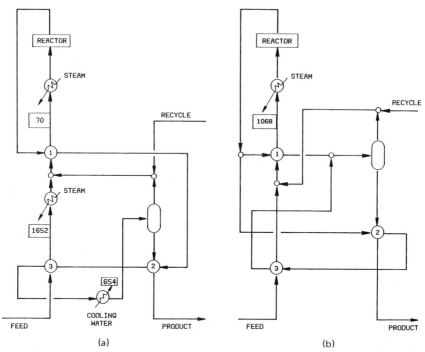

Fig. 11-1   (a) Traditional and (b) alternative heat exchanger design for the front end of a specialty chemical plant: (a) six heat exchangers, 1722 total units of heat supplied, 654 total units of cooling water used; (b) four heat exchangers, 1068 total units of heat supplied, no cooling water used.

Both of these systems have limitations. Because of the large number of possible solutions, mathematical search techniques generally require a large number of simplifying assumptions to make solutions practical. In addition, the flexibility necessary to install stream splitting and parallel paths is generally limited. For both methods a fixed minimum temperature difference in any heat exchanger must generally be assumed. In addition, the systems are generally inflexible on feedback. This limits the opportunity to identify small changes in process conditions that might lead to even more optimum arrangements. Thus, most investigators have labeled the solutions published for specific heat network problems as "near optimum."

Most recent work has been based on the "rules-of-thumb" approach. The simplest and most readily understood rules are those advanced by Ponton and Donaldson.[2] These investigators simply recommend repeated matching of the hot stream with the highest supply temperature against the cold stream that is to be heated to the highest target temperature in an approach to a design solution. The approach is quick, easily understood, and often leads to quite good networks. However, much better solutions have been found to certain problems using other methods. Boland and Linnhoff[1] describe a better solution to the so-called "7 SP 1" problem than that arrived at by Ponton and Donaldson.

B. Imperial Chemical Industries – Leeds Approach

Considerable research was done at Leeds University by Linnhoff, Flower, and their co-workers. This was later extended in the ICI Corporate Laboratory where Linnhoff went to work after receiving his degree. These investigators attempted to substitute thermodynamic analysis for rules of thumb and to develop a systematic approach to the synthesis of a solution. They relied on this analysis to help predict boundary conditions and simplify design problems. Lamaeda and co-workers from Chioda Chemical Engineering and Construction Company have also worked in this general area and have contributed to the application of this general type of technique.[3] Although many others have also contributed, we shall draw on the published work of these two groups in an attempt to illustrate the principles of the approach without going into a great deal of detail.

The thermodynamic limit to heat recovery can be calculated from an overall heat balance. The difference in the enthalpy changes of the hot and cold process streams must equal the difference between the enthalpy of the heat supplied from hot utilities and the heat rejected to cold utilities. The amount of heat that can actually be recovered by integration, however, depends in a very complex way on the temperatures and heat capacity flow rates (MC_p) of the various streams involved. Consider the array shown in Table 11-1, in which two cold streams are to be heated up to final tempera-

Table 11-1

Illustrative Example of a Heat Exchange Network

Stream number	Stream type	Heat capacity flow rate MC_p (kW/°C)	Starting temperature T_s (°C)	Final temperature T_f (°C)	Heat load $MC_p(T_s - T_f)$ (kW)
1	Cold	3.0	60	180	−360
2	Hot	2.0	180	40	280
3	Cold	2.6	30	105	−195
4	Hot	4.0	150	40	440
Total					165

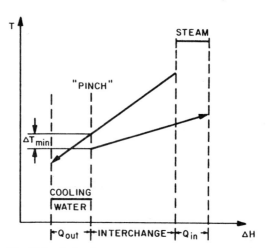

Fig. 11-2 A simple two-stream heat recovery network.

tures of 180°C and 105°C, respectively, and two hot streams need to be cooled from 180°C and 150°C to 40°C. Because more heat is available from the hot streams than is needed by the cold streams, one might assume that this process could be accomplished without the input of external heat. This is not the case, however, because of the temperature relationships and the slopes of the heat capacity flow rate curves.

The logic for this statement is shown for a simple two-stream problem in Fig. 11-2. Here the temperature–enthalpy $(T-Q)$ curves for one hot and one cold stream are plotted. A minimum driving force for heat exchange is specified, which results in an overlap at one end that must be removed with cooling water and one at the other end that must be provided by an outside heat source. In this case steam was identified as the utility required.

For more than one pair of streams, a composite curve must be established for all hot streams and all cold streams. This can be accomplished either mathematically or graphically.[4] This procedure is shown in Fig. 11-3a, in which the combined $T-Q$ curves for two streams A and B are generated by simple graphical methods. The two curves are drawn by placing the origin of the higher-temperature curve at the heat transferred point that corresponds to the end of the lower-temperature curve. The combined curve in the region where the individual curves overlap is formed by simply connecting the opposite corners of the overlapping region, as shown in Fig. 11-3b. The resulting curve is shown in Fig. 11-3c.

The combined curve can also be constructed mathematically by adding the heat capacity flow rates for the two curves and plotting the result as the slope in the region of temperature overlap. For example, for the two hot

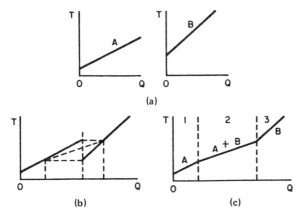

Fig. 11-3 Procedure for constructing composite $T-Q$ curves: (a) provide individual heat availability lines; (b) combine lines in the same temperature range (draw a diagonal line); (c) complete the combined $T-Q$ curve.

streams listed in Table 11-1, the sum of the heat capacity flow rates is 6.0 kW/°C. The streams overlap in the temperature range of 40 to 150°C. Because the temperature is the ordinate of the curve we want, if one plots the reciprocal of the sum of the heat capacity flow rates, namely, $\frac{1}{6}$ as the slope of the curve between 40 and 150°C, one arrives at an identical curve to that developed by the graphical method.

The $T-Q$ curves for the problem defined in Table 11-1 are shown in Fig. 11-4 with the assumption that a minimum temperature difference for heat transfer of 10°C will be held at the pinch. As shown in this diagram, approximately 60 kW plus the 165 kW surplus identified in Table 11-1 must be removed by cooling water to maintain the 10°C minimum approach temperature. From this graph one can quickly determine the impact on the amount of heat required of reducing the 10°C minimum approach temperature to 5°C. This is done by moving the lower curve to the left (closer to the upper curve) until the vertical distance (ΔT) at the pinch is 5°C. The dotted line shows that the heat required would be reduced to about 45 kW if the minimum temperature difference were reduced to 5°C.

These composite $T-Q$ curves identify the pinch in a heat exchanger network. The pinch is an important point, as will be demonstrated later in the discussion. In addition, the curves show the amount of available heat that can be recovered as a function of the approach temperature assumed. If the approach temperature is a very critical variable, this can be quickly analyzed by the designer before beginning laborious heat exchanger design calculations. A curve similar to Fig. 11-4 can be plotted up for each system to be studied. One of the interesting facets of the system described by Fig.

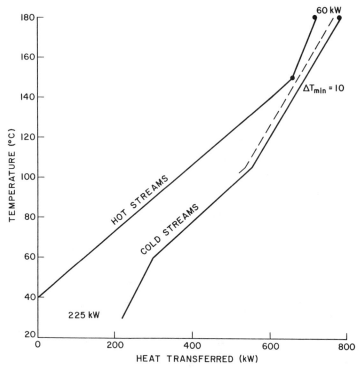

Fig. 11-4 Composite $T-Q$ curves for the example of Table 11-1.

11-4 is that even with zero temperature difference at the pinch, some heat must be supplied from external sources to meet the target temperature set for the cold stream. In this case approximately 25 kW are required.

C. Minimum Number of Units Equals Minimum Cost

In the design of a heat exchange system, the heat transfer coefficients of the exchangers are almost independent of matching decisions. This is also generally true of driving forces once the minimum temperature difference is set, unless very sharp changes in the physical properties of the streams are encountered at the temperature levels being considered. This might be an important consideration in polymer systems. Because the coefficients are relatively constant, the total surface area required in a given system can be simply calculated from the overall equation

$$Q = UA \, \Delta T, \tag{11-1}$$

where U is an overall heat transfer coefficient (Btu/h · ft²°F). For the problem under consideration, the total heat transferred in the system is approximately 780 kW. For this amount of heat it can be expected that the overall surface area of different networks used to accomplish the same quantity of heat interchange is going to be about the same. Indeed, Hohmann reports that this is usually true within ±3%.[5]

Heat exchange surfaces usually have exponential cost equations, i.e.,

$$\text{cost} = A(\text{surface area})^B, \qquad (11\text{-}2)$$

where B is less than 1.0. Exponent B is generally on the order of 0.6. When multiple exchangers are required for a single service, exponent B will approach 1.0 for the exchangers alone. Because piping and instrumentation are still very much simpler with fewer exchangers, the *total* cost exponent will always be less than 1.0. We can conclude, therefore, that if the total surface area is approximately constant but the number of heat exchange services is reduced, then the economies of scale will serve to make heat exchange networks that require fewer units less expensive.

This being the case, designers are very interested in the minimum number of heat exchange services that can be used, because this corresponds to the minimum cost. Bolland and Linnhoff developed the minimum number of units for a simple system as follows:

$$Z_{\min} = N_{\text{proc}} + N_{\text{util}} - 1, \qquad (11\text{-}3)$$

where N_{proc} is the number of process streams to be heated or cooled and N_{util} the number of utility streams to be dealt with. In the case being considered the number of utility streams would be two: steam and cooling water. Investigators go on to point out that a general relationship can be derived based on Euler's network relationship, resulting in the following equation:

$$Z_{\min} = N_{\text{proc}} + N_{\text{util}} - S + L, \qquad (11\text{-}4)$$

where S represents the general case of more than one branch in the network and L identifies the number of internal loops in the network. Branches and loops develop as more complex networks are synthesized in seeking the optimum. If a network were developed in which two separate branches were conceived, one for the hot and cold streams with equal heat load and one for the remainder of the problem, S would be two. With two branches, if a single loop were created among several heat exchangers in the network, the minimum number of units would then correspond to the initial estimate of the minimum number of units in the heat exchange system given by Eq. (11-3).

For the demonstration problem, 780 kW of heat must be exchanged at a minimum temperature difference of 10°C. In addition, for a single-branch network with no loops, the minimum number of heat exchange units will be $4 + 2 - 1$ or 5 units. Branching may reduce this.

D. Network Diagrams

Linnhoff *et al.* developed a system for representing heat exchange networks that facilitates solutions. In this system hot streams flow from left to right at the top of the diagram and cold streams flow from right to left at the bottom. The graphical representation is then divided into subnetworks by temperature levels based on judgment, and heat exchange matches are shown by drawing a circle in each of the streams to be matched connected by a vertical line. Other exchangers using external utilities are marked by circles labeled H or C. One starting layout for the subnetworks of the streams in the demonstration example is shown in Fig. 11-5. We can now begin the process of synthesizing a heat exchanger network that represents minimum utility requirements and minimum capital costs. The objectives are to use no more than 60 kW of outside steam, reject no more than 225 kW to cooling water, and install only 5 heat exchange units.

The process is still an iterative one. Various selections are made and tested against the objectives for minimum utility requirements and minimum capital costs. One starting point could be the lineup shown in Fig. 11-6, in which the 60 kW of heat are put into stream ☐1 and the rest of stream ☐1 is heated up by interchange. As can be seen, this solution offers an unsatisfactory approach to the objectives set out earlier. More (105 kW) external heat (those circles marked H) and 10 separate heat exchange services are required. This first-pass network does not achieve our stated objectives, because some exchanger choices created, at an early stage in the synthesis, a

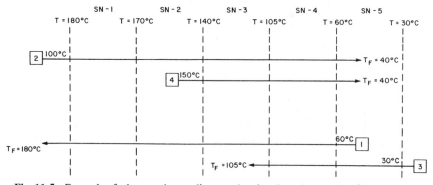

Fig. 11-5 Example of a heat exchange diagram showing the subnetworks of the streams.

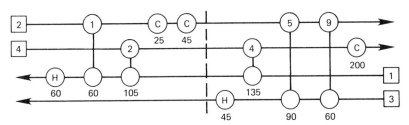

Fig. 11-6 First-pass solution for the heat exchange network of Fig. 11-5, including heat flows in kilowatts.

situation that resulted in constraints later on in the network. These early choices prejudiced the freedom of later choices. Freedom of choice can be maximized throughout the synthesis by matching a hot stream to the particular cold stream in that section that has the highest temperature. This is the main function of the temperature partitions between the subnetworks set up in the original table. The intent is to pass heat out of a given subsection to the one with the next lowest temperature only after all the cold streams within the first section have been provided for.

To develop the synthesis in a more profitable direction, ways of substituting heat interchanges between hot and cold streams for the additional steam heaters required in the original attempt at a solution must be found. The cooler on stream ② at 45 kW can be the heater on stream ③ at the same size. This eliminates the need for 45 kW from steam, reduces the lost work in the exchanger (ΔT), and reduces the number of heat exchangers to 9. Combining heat exchangers ⑤ and ⑨ on streams ② and ③ and exchangers ② and ④ on streams ④ and ① leaves the result shown in Fig. 11-7, which is almost a match for the initial objectives. Note that the minimum temperature difference for heat transfer of 10°C is maintained at the closest approach of all heat exchangers.

The name applied by Linnhoff and Flower[6] to this method is the temperature interval method. Other investigators have used similar systems in the ordering of all heat loads to arrive at the same type of solution. The methods differ at the point in the process when stream splitting is employed in complex networks to approach the optimum configuration. Note that judgment is still required in progressing from the first attempts to the final version. However, the system of stream layout very much facilitates the matching of appropriate duties and the generation of alternate solutions. This is very important in more complex problems and may even be pertinent in this sample problem. The potential to branch the network and further reduce the number of exchangers remains to be explored.

More sophisticated networks can benefit from a more organized evolu-

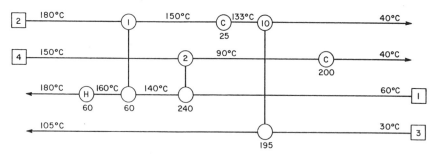

Fig. 11-7   A better solution for the heat exchange network of Fig. 11-5, including heat flows in kilowatts.

tionary procedure, which Linnhoff and Flower[6] branded the evolutionary development method. This method concentrates on evaluating the impact of any heat exchange decision on the freedom of choice remaining for other exchangers. Operating and safety constraints can be included. Rules are provided to guide the direction of the evolution, and they become quite complex in complex situations. Numerous complex networks are solved by the authors with this approach, giving more optimal solutions than those advanced by previous investigators. The reader is referred to Reference 6.

E. Further Optimization

It is easy to see that the network proposed in Fig. 11-7 is better than the first-pass proposal (Fig. 11-6). Energy recovery is better and fewer exchangers are used. However, other more nearly optimum networks that do not violate any of the design or thermodynamic constraints are also possible for this problem. Study of the diagram can identify them. For example, heat exchangers ① and ② both operate at minimum temperature differences at adjacent ends of the exchangers. Any increase in temperature difference at these points (an increase in irreversibility) would cause the need for more steam at H . This is not a possibility. However, the position of the cooler on stream ② before exchanger ⑩ prompts further investigation. Switching the cooler and exchanger ⑩ is the first thought. Such a solution is shown in Fig. 11-8a. It appears feasible from a temperature standpoint. However, the size of the cooler would increase markedly because its driving force is greatly reduced. Exchanger ⑩ would change only slightly in size, because the percentage change in the LMTD is small. The economics of the two configurations would have to be tested in more detail.

Another approach that would achieve the same goal (position all the cooling at the end) would be to split the heating of stream ① between streams ④ and ②. This places all the cooling in one exchanger at the

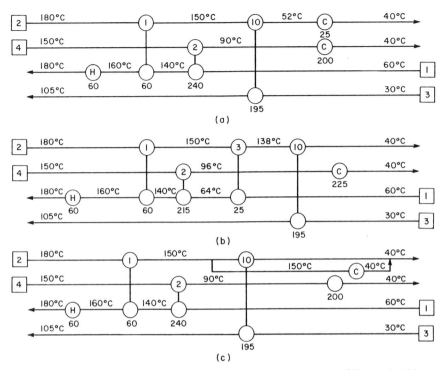

Fig. 11-8 Further alternate solutions for the heat exchange network of Fig. 11-5 (all heat flows in kilowatts): (a) switching a cooler and exchanger ⑩ ; (b) dividing the heating of stream 1 between streams ④ and ②; (c) a more sophisticated stream splitting approach.

expense of another interchanger (see Fig. 11-8b). Again, this is feasible from a temperature standpoint and approaches the temperature driving force profile of the original solution. The economics will depend on the complexity of the piping around the interchanger versus the simpler "one exchanger for the whole job" solution.

An alternative using the more sophisticated approach of stream splitting is shown in Fig. 11-8c. This technique is often useful in more complex problems. The heat capacity flow is adjusted to give the desired outlet temperatures. Again, the mean temperature difference on the cooler will change markedly because of the lower outlet temperature.

The purpose of this is to demonstrate that many near-optimum solutions can be developed, the economics of which need to be tested. In more complex systems, the problem of finding the best solution is far from trivial even if other considerations such as safety and operability are not initially considered. Obviously, any practical design must also meet these additional criteria.

F. The Problem Table

For more complicated problems, it would be a lengthy procedure to evaluate possible designs for each subnetwork and discuss their suitability by the means of sketches like those in Figs. 11-6–11-8. The problem table shown in Table 11-2[6] helps to carry out the search in a systematic way. The problem and its subnetworks are restated on the left-hand side of the table. The "deficit" column shows the difference between the heat input by any heater and the heat output from any cooler. This is the net heat requirement for any subnetwork and is defined by the following heat balance: $D_i = i_i - O_i$, where i is the heat input to a given subnetwork and O the heat output from a given subnetwork. If no utility heat is supplied to any of the subnetworks, then all of the surplus heat from any subnetwork is passed along to the next-lower-temperature subnetwork as a heat input. Thus, we can calculate the output from the next subnetwork in the line as follows:

$$O_{i+1} = O_i - D_{i+1}. \tag{11-5}$$

The calculation of the deficits ($MC_p \, \Delta T$) in Table 11-2 proceeds as follows (all values in kilowatts):

| Subnetwork | Stream | Change in process (°C) | | Deficit |
		From	To	
1	Heating ☐1	170	180	$3 \times 10 =$ 30
2	Cooling ☐2	180	150	$2 \times (-30) =$ -60
	Heating ☐1	140	170	$3 \times 30 =$ 90
	Total			30
3	Cooling ☐2	150	115	$2 \times (-35) =$ -70
	Cooling ☐4	150	115	$4 \times (-35) = -140$
	Heating ☐1	105	140	$3 \times 35 =$ 105
	Total			-105
4	Cooling ☐2	115	70	$2 \times (-45) =$ -90
	Cooling ☐4	115	70	$4 \times (-45) = -180$
	Heating ☐1	60	105	$3 \times 45 =$ 135
	Heating ☐3	60	105	$2.6 \times 45 =$ 117
	Total			-18
5	Cooling ☐2	70	40	$2 \times (-30) =$ -60
	Cooling ☐4	70	40	$4 \times (-30) = -120$
	Heating ☐3	30	60	$2.6 \times 30 =$ 78
	Total			-102

Table 11-2

Problem Table for the Heat Exchange Network of Fig. 11-5

	Temperature (°C)		Stream present in subnetwork?				Deficit[a] (kW)	Accumulated flow		Maximum permissible flow	
			Cold streams		Hot streams			Input (kW)	Output (kW)	Input (kW)	Output (kW)
Subnetwork			1	3	2	4					
SN-1	180	170	×		×		30	0	−30	**60**	30
SN-2	150	140	×		×	×	30	−30	−60	30	0
SN-3	115	105	×		×	×	−105	−60	45	0	105
SN-4	70	60	×	×	×	×	−18	45	63	105	125
SN-5	40	30			×	×	×	−102	63	**165**	125	**225**

[a] The deficit is equal to the heat input from the hot stream minus the heat demand of the cold stream.

These numbers are then inserted in the problem table opposite their respective subnetwork rows. Calculation of the accumulated inputs and outputs is done as indicated in Eq. 11-5. The deficit of 30 kW for SN-1 becomes an output of -30 kW. No heat is transferred into SN-1 from any of the other networks, so its accumulated input is zero. The input to SN-2 from SN-1 is -30 kW. Its output is calculated as follows: $O_2 = -30 - D_2 = (-30) - 30 = -60$ kW. The input to SN-3 from SN-2 is -60 kW, and its output is $O_3 = -60 - (-105) = 45$ kW. For SN-4, the output is $45 - (-18) = 63$ kW, and for SN-5 the input is 63 kW and the output is 165 kW.

You will recognize the 165 kW as the total of surplus heat from Table 11-1. This represents the net heat that would have to be rejected to the cooling medium if no utility heat were required to solve the heat exchange network. If any of the values under "output" are negative, as is the case here for SN-1 and SN-2, process utility heat must be introduced to the subnetworks to increase their outputs to zero. Thermodynamically, it makes sense to introduce this utility heat at the highest available temperature. The maximum amount of utility heat required is determined from the largest negative value under "output." The real objectives for each subnetwork are then recalculated by the same procedures to arrive at the last two columns in the problem table. Sixty kilowatts of heat are introduced as an input for SN-1, changing its output from -30 to 30 kW. Following the calculation procedures discussed earlier, this reduces the output of SN-2 to zero, the desired minimum. Moving down the table, we arrive at an output from SN-5 of 225 kW, the same cooling load calculated graphically earlier in the discussion of this problem.

The completed problem table (Table 11-2) is a useful way of organizing more complex heat exchange network problems. The table shows

(1)   values for the total process heat and cooling loads that will be required if maximum energy recovery is achieved;

(2)   the maximum permissable figures for the heating and cooling loads of each subnetwork, which must not be exceeded if the final network is to be optimum from an energy recovery point of view.

The data presented in Table 11-2 do not depend on how the subnetworks are constructed. Thus, an optimum selection of subnetworks is not a prerequisite to developing the optimum heat exchanger network.

The example discussed here is meant to illustrate the principles of the method and their relationship to thermodynamics. As before, automatic solutions to heat exchanger network problems are not guaranteed. In addition, complexities are introduced by the need to deal with changes of phase in various heat exchangers, which introduce the possibility for internal temperature driving force pinches. One is still faced with the choice of different sources of process heating and cooling, and these must be introduced at the proper point in the network. Failure to do this could impose some constraints on choices later on in the network, which has already been shown to lead to less than optimum solutions. Various methods of systematizing subnetwork selection and exchanger matching have been proposed to facilitate this process. Additional constraints for safety and operability are also discussed by Linnhoff and Flower.[6]

To provide the reader some practice with the method, a second network problem is outlined here. Develop the solution using the problem table approach that resulted in Table 11-2 for the earlier example. The completed table is presented so your results can be checked.

Problem: Develop a Heat Exchange Network for the Following Streams

Stream number	Stream type	Heat capacity flow rate MC_p (kW/°C)	T_s (°C)	T_f (°C)	Heat load $C_p(T_s - T_f)$ (kW)
SN-1	Cold	3.0	60	180	-360
SN-2	Hot	2.0	180	40	280
SN-3	Cold	2.6	30	130	-260
SN-4	Hot	4.0	150	40	140
Total					100

Solution

Stream number	Temperature (°C)		Steam present in subnetwork?				Deficit (kW)	Accumulated flow		Maximum permissible flow	
			Cold streams		Hot streams			Input (kW)	Output (kW)	Input (kW)	Output (kW)
			1	3	2	4					
		180	×								
SN-1	180	170	×				30	0	−30	−60	30
SN-2	150	140	×		×		30	−30	−60	30	0
SN-3	140	130	×	×	×	×	−30	−60	−30	0	30
SN-4	70	60	×	×	×	×	−28	−30	−2	30	58
SN-5	40	30		×	×	×	−102	−2	100	58	160

G. Relation to Available Energy (Exergy)

Umeda, Harada, and Shiroko[7] point out that the $T-Q$ curves shown in Fig. 11-3 can be replotted to show the available energy requirements in place of enthalpy requirements. This is done by plotting the quantity $T - T_0/T$ as the ordinant as opposed to the simple temperature. The concept of a pinch point is again applied to identify the minimum available energy input and rejection for any good heat exchanger network. The composite hot and cold stream curves are constructed as before.

Figure 11-9 shows the available energy diagram for the demonstration problem discussed earlier. In this diagram the area under the curve represents the available energy corresponding to the heat contained up to any point in question. Thus, the total available energy of the composite hot streams is the area under the curve from the origin to point B. That wasted in the cooling water is the section from the origin to point A. Conversely, the available energy recovered in the cold streams is the area of the curve for the composite cold stream from its origin up to point B'. The additional available energy introduced into the cold stream comes from the utility, and it is represented by the area under the curve from point B' to point C. Specifying different limitations for the minimum temperature difference for heat transfer would have the effect of moving the lower curve to the right or left along the Q axis. With these plots, which are simply derived from the $T-Q$ curves, the design engineer can quickly ascertain the impact on available energy (potential fuel savings) of his constraints on the minimum temperature difference in the heat exchanger network. Very roughly, the area under the curve from point B' to point C is about 20 kW. The amount

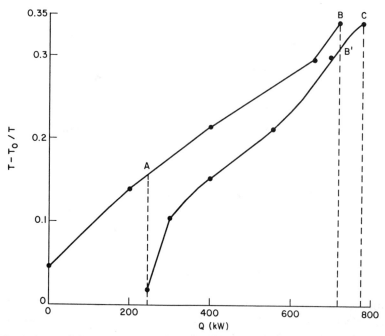

Fig. 11-9   Available energy versus heat Q for the heat exchange network of Fig. 11-5.

of available energy expended to supply this 20 kW to the process depends on the temperature level of the steam supplied. Similarly, the amount of available energy wasted to the cooling water is about 24 kW. This is only about 10% of the total enthalpy wasted because of the relatively low temperature of the streams from which the heat is rejected. Qualitatively, reducing the cooling water load by some fraction of the 60 kW of steam heating required does not have a large impact on the amount of available energy wasted to cooling water.

H. Summary of the Method

Very briefly, we can summarize the methods of process synthesis as follows.

(1)   Identify the minimum number of units and identify the minimum utility requirement by creating a problem table.

(2)   Synthesize a maximum energy recovery network using the subnetworks and the problem table.

(3)   Reduce the number of individual heat exchange units by replacing heaters and coolers with exchangers.

(4)   When a network is close to the minimum number of units, promote features of practical interest including safety and operability constraints.

(5)   Compare alternatives that approach the optimum solution economically.

II. Applications to Cogeneration Systems

A. The Significance of the "Pinch"

Townsend and Linnhoff[8] showed how the analysis of heat flow characteristics identifies the point in a process at which the energy bottleneck (or pinch) occurs. This pinch divides any process into thermodynamically independent parts. The division is useful in considering both power system integration and heat exchanger network analysis. Full benefits from the advantages of cogeneration would be obtained only if the power component of the system is properly placed relative to the pinch. Referring back to the composite curves in Fig. 11-4, we note that heating is required at higher temperatures than the pinch and cooling at lower temperatures. Above the pinch all of the available heat in the composite hot stream can be utilized by the cold stream in that section of the diagram. Similarly, below the pinch all of the cooling capability of the composite cold stream can be utilized by interchange with the hot stream. If a small element of heat dQ is transferred from the hot side of the pinch to the cold, then the same amount of heat must be supplied from the external heat source to the cold stream at temperatures above the pinch to restore enthalpy balance. This cannot come from the composite hot stream at temperatures lower than the pinch, because this would involve transfer at a less than minimum (or worse, a thermodynamically impossible) driving force. So, if the hot utility is increased by dQ, then the cold utility must be increased by dQ as well. Thus, for minimum energy solutions, the golden rule is do not transfer heat across the pinch.

B. The Concept of Appropriate Placement

The foregoing discussion provides the clue to where the integrated heat engine must be placed to achieve full utilization of the cogeneration potential of the system. *The exhaust heat must be supplied at temperatures higher than the pinch.* Any exhaust heat that is transferred at temperatures lower than the pinch cannot replace the hot utility. Thus, the appropriate placement for the heat engine that rejects its exhaust heat to the heat exchange network is at temperatures higher than the pinch. If the exhaust heat is rejected below the pinch, the synergism of the cogeneration system is lost. Please note that an engine that both *absorbs and rejects* heat at temperatures lower than the pinch is also appropriately placed, because it does not disturb the heat balance across the pinch.

This concept can also be applied to heat pumps. The heat pump is basically a device that absorbs heat at the lower temperature, consumes work, and rejects the heat plus the work at a higher temperature level. For such a device to save energy, it must absorb heat at temperatures lower than the pinch and reject it at temperatures higher than the pinch. Therefore, to be appropriately placed it must pump heat across the pinch. In other positions the effect of the heat pump is simply additional cooling load. More details of this analysis are presented in Reference 8.

III. Thermoeconomics

As discussed throughout this book, process design and improvement consists of the art of forging an attractive solution for a problem that involves both scientific and economic constraints. It has also been noted that what seems to be the optimum solution in one area of a process plant can often have interactions with other parts of the process or utility system that cause the sum of the apparently optimum parts to result in a nonoptimum whole. To the extent that these nonoptimalities are within the range of accuracy of process design tools, they are insignificant. However, when opportunities to reduce both the energy consumption and the capital cost of a plant design are missed, they can become very significant indeed. To complicate this problem further, the process designer rarely has available, *at the time it is needed,* all of the economic data required to make a thoroughly intelligent decision. This happens because capital cost estimates take time and manpower and cost money. In addition, the data required by the cost engineer often includes the process designer's final product, rather than the rudimentary conceptual thinking. Thus, we have the classic circular dilemma: good cost estimates are not available until the process design is complete, and a process design is not complete until some cost estimating is done for optimizing the design.

A number of investigators[9] have attempted to devise systems that combine thermodynamic and economic considerations in an effort to bring the process designer all of the necessary information in a timely fashion. These efforts are generally termed "thermoeconomics."

A. The Principle of Thermoeconomics

The general approach is to carry out a second law analysis or an entropy balance for the process to go along with the usual heat and material balance. This might take the form of an available energy (availability, exergy, essergy) balance, or perhaps an entropy balance is carried out. Whatever its form, this analysis identifies places in the process where irreversibilities exist and allows the quantification of losses of work (availability, exergy, essergy) that result from these irreversibilities. The economic environment is introduced by establishing prices for these losses at various places in the process. In places where the losses can be traced directly to fuel or purchased electricity, the prices for the available energy lost or degraded can be determined in a rather straightforward manner. In places where several possible sources of available energy contribute to the energy being dissipated, an average price is generally used. Prices may vary from zone to zone in a process because of inherent losses or releases associated with the chemistry or the processing equipment.

It is essential that these prices for available energy include an appropriate way to deal with the investment required in the plant to provide the available energy in the form used by the process. To do otherwise only brings half of the economic environment into play. As discussed in Chapter 7, the process is broken down into segments and, ultimately, to individual pieces of equipment if necessary, so that a reasonably detailed accounting can be made of the thermoeconomic fluxes. For each process segment or piece of equipment, the available energy balance can be written as

$$A_{in} - A_{out} = A_{lost} = T_0 \Delta S. \qquad (11\text{-}6)$$

This is the rate of entropy formation, the "lost work," or the rate of available energy (or exergy or essergy) dissipation, depending on terminology. If α_i is the cost of available energy in segment i, then the thermoeconomic flux associated with the dissipation in segment i is[10]

$$V_i = \alpha_i T_0 \Delta S_i. \qquad (11\text{-}7)$$

If prices are different at the inlet and outlet of a segment, the thermoeconomic dissipation flux becomes

$$V_i = \alpha_i A_{in_i} - \alpha_0 A_{out_i}. \qquad (11\text{-}8)$$

This amounts to a money balance across the process segment. Obviously, a simplified nomenclature is required for complex problems.

The next step is to prepare an accounting of the fluxes for the process being studied. The thermoeconomic dissipation fluxes in each segment are used to indicate the areas with the most potential for improvement, just as the lost work breakdowns were used in thermodynamic analysis.

Before going on to an example, a word about prices is essential. If available energy is dissipated in the flow of a liquid through a pipe, the price of that dissipation can be calculated from the price of the work supplied to the pump that moves the liquid in dollars per kilowatt. Alternatively, all prices could be stated as a ratio of the fuel price.

The capital component of the cost of pumping can be derived from a knowledge of the capital cost of the pump and its power supply plus the current depreciation rates and desired return on investment. The total price is the sum of the available energy and fixed parts:

$$\alpha_i = \alpha_A + \alpha_F; \tag{11-9}$$

$$\alpha_i = \alpha_A + \alpha_C + \alpha_L, \tag{11-10}$$

where α_C is the capital related cost and α_L the labor cost;

$$\alpha_i = \alpha_A + C_i(R_D + R_R) + \alpha_L, \tag{11-11}$$

where C is the capital cost of the pump, R_D the depreciation and maintenance fraction per unit time, and R_R the return requirement per unit time. Thus, if a pump and power supply cost $100,000 total, R_D were 0.1, and a 15% simple return on investment were required, the capital fraction of the cost of supplying the horsepower needed for any process segment in the simplest terms would be

$$\alpha_C = \frac{(100,000)(0.1 + 0.15)}{\text{design power}} = \frac{25,000}{\text{design power}}. \tag{11-12}$$

More complex allocations of the capital cost as a function of design power can also be used, if appropriate.

For cases in which process conditions vary appreciably from design, it is advisable to keep available energy and capital (fixed) costs separate to avoid mixing variable and constant costs into a single price. A simple illustration should clarify the concept. High-pressure steam is often used to drive a turbine attached to a compressor, generator, or other process machinery. The high-pressure steam is converted to other products by the turbine, i.e., power and lower-pressure steam. Assuming the steam comes directly from a boiler, the price of the steam is

$$\alpha_{HP} = \text{(cost of fuel, water, chemicals)}/M_{HP}$$
$$+ \, [\text{(boiler capital)}(R_D + R_R) + \text{(labor cost)}]/M_{HP}, \quad (11\text{-}13)$$

where M_{HP} is the net production of high-pressure steam in pounds per unit of time. Assuming the boiler efficiency does not change over the range of interest, the first term is a constant. The second term will vary as a function of throughput *unless* alternate customers are found for any spare capacity. In a reasonably well-balanced steam system, a certain percentage of spare capacity is often built into the fixed cost term, thus distributing spare capacity costs over all steam production.

Accounting data are normally available to make the calculation of α_{HP} straightforward. What are the prices of the power and low-pressure steam products from the turbine? A money balance fixes one constraint:

$$\alpha_{HP} M_{HP} + \text{fixed cost (turbine)} = \alpha_{LP} M_{LP} + \alpha_w W, \quad (11\text{-}14)$$

where W is the power extracted in the turbine and α_w and α_{LP} are the unknown prices. Generally, the capital cost of the turbine (and condenser, coolant supply, etc., if applicable) and the amounts of power and low-pressure steam are known. This leaves two unknowns, so another equation is needed.

This other equation will depend on a philosophical (management) judgment. The price for available energy can be assumed constant for both the low- and high-pressure steam, as in the availability pricing approach of Chapter 3. Thus,

$$\alpha_{HP} = a A_{HP}, \quad (11\text{-}15)$$

$$\alpha_{LP} = a A_{LP}; \quad (11\text{-}16)$$

dividing, we obtain

$$\alpha_{LP} = \alpha_{HP}(A_{LP}/A_{HP}). \quad (11\text{-}17)$$

Substitution in Eq. (11-14) gives

$$\alpha_{HP} M_{HP} + [\text{fixed cost (turbine)}] = \alpha_{HP}(A_{LP}/A_{HP}) + \alpha_w W, \quad (11\text{-}18)$$

allowing calculation of α_w.

Another approach would be to consider the power as a byproduct. In this case the steam would be priced as would that from a hypothetical low-pressure boiler (following the outline for the high-pressure steam). Subtracting this value from the left-hand side of Eq. (11-14) gives a different, and lower, price for the power generated.

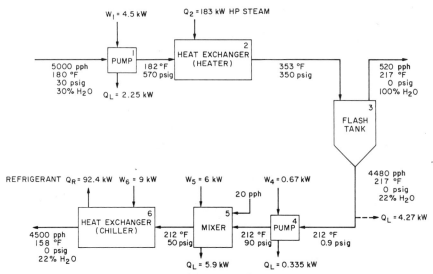

Fig. 11-10 Sample mass and energy flow diagram for thermodynamic accounting. $\Sigma\ w_{in} =$ 20.17 kW, $\Sigma\ Q_0 = 12.75$ kW. Q_R, heat removed by refrigerant; Q_L, heat lost at any step; Q_n, heat input at step n. From Reference 11.

Another alternative would be to price the power as the equivalent of purchased electricity. Then the price of low-pressure steam would be calculated by subtraction. Other approaches are also possible. The point is that when more than one product is produced in a process segment, unique pricing calculations are not possible from thermodynamic analysis and accounting data alone.

Tribus and El-Sayed[11] developed a more complex example to illustrate thermoeconomic accounting. The process is shown in Fig. 11-10. Each piece of equipment is a separate process segment. A solution of 30% sodium palmitate is pumped to high pressure, heated, flashed to a low pressure to remove excess water, mixed with an additive, and cooled in a chiller. Small power inputs and heat losses are encountered throughout. T_0 is taken to be 528°R.

The available energy balance is shown in Fig. 11-11. Power inputs are pure available energy and heat flows are converted to available energy by the Carnot factor $1 - T_0/T$. Dissipations are the difference between inlet and outlet exergies as shown in Eq. (11-6). For example, in the first segment, the pump, inputs are 16.68 kW in the feed and 4.5 kW in electric power for a total of 21.18 kW. Outputs are 19.29 kW in the product and 0.39 kW in the

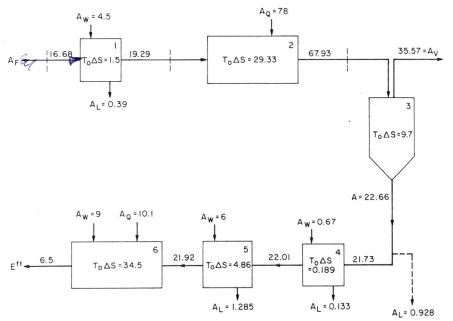

Fig. 11-11 Available energy flow diagram for the process in Fig. 11-10. All flows are in kilowatts. $\Sigma A_{in} = 124.95$ kW, $\Sigma A_{out} = 44.315$ kW, $\Sigma T_0 \Delta S = 80.08$ kW. A_F, available energy of the feed; A_w, available energy input from electricity; A_Q, available energy input from heat; A_L, available energy lost at any step; A_V, available energy lost in the vapor. From Reference 11.

heat lost from the pump for a total of 19.68 kW. The difference is 1.5 kW, the dissipation in the pump itself. Note that in the chiller, the power required in the refrigeration system is expended *on the system* and is shown as an input along with the power for the motor.

The prices for available energy in various forms vary appreciably:

Purchased power	4.37 ¢/kW · h
High-pressure steam	3.2 ¢/kW · h
Low-pressure steam	3.0 ¢/kW · h
Refrigerated brine	17.4 ¢/kW · h cooling load (including a capital factor for the brine system)

The thermoeconomic balances are shown in Fig. 11-12. The data are given in cents per hour for inputs, losses, and dissipation. The fluxes were

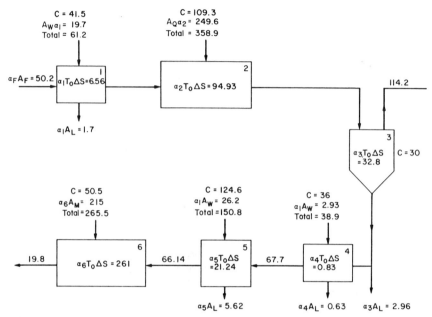

Fig. 11-12 Thermoeconomic flow diagram for the process in Fig. 11-10. All flows are in cents per hour. $\Sigma_i A_i \alpha_i = 564$¢/h, the total capital cost is 392¢/h (feed plus available energy input). From Reference 11.

calculated by multiplying the prices for the various available energy inputs by the lost work at each step. The combined thermodynamic–economic impact identifies the major process dollar losses. In order of importance, they are the chiller, heater, flash drum, and mixer. The process engineer must still make the creative leap to generate new ideas, but he is now guided to those areas that are important economically as well as thermodynamically in a single step. Thus, more systematic practical guidance for process and design improvement efforts is provided.

There is still the need for reasonably accurate equipment cost data early in the design phase. These need to be internally consistent if not absolutely accurate. This means that the relative costs of different types and sizes of equipment need to be correct on a total erected cost basis, but that corrections for inflation, location, type of contract, etc., may not be needed. This is not a trivial task. In addition, cost engineers have the tendency to renege on preliminary numbers once the final design is available on which to base a more detailed estimate. However, many companies maintain simplified estimating systems, and a number of technical magazines also publish data on a consistent enough basis for use in studies of this type.

B. Applications of Thermoeconomics to Process Improvement

Referring to the prices for available energy given earlier, the dissipations shown in Fig. 11-12 were calculated by identifying the source of available energy when it was practical and averaging prices when it was not. For example, in segment 1, all of the available energy is introduced by electric power purchased at 4.4 ¢/kW · h. Thus, both the dissipation and the losses are valued at this price. This is also true in segments 4 and 5.

Segment 6 uses a mixture of electric power and refrigerated brine. The average price of the available energy supplied is

$$\alpha_6 = [(9)(4.37) + (10.1)(17.14)]/(19.1) = 11.26. \qquad (11\text{-}19)$$

A few of the other prices require more complex analysis.[11]

The additional factor required to complete the dollar flows is to add in the investment cost. The following capital costs (in thousands of dollars) and operating life (in years) were used by the authors:

Zone	Capital cost ($1000)	Operating life
1 Pump	15	14
2 Exchanger	46	25
3 Flash tank	12	20
4 Pump	13	14
5 Mixer	45	14
6 Exchanger	18	14

With straight-line depreciation and a 15% simple return on investment, the following capital cost dollar fluxes can be added to the available energy fluxes, assuming 8000-h/yr operation:

Zone	Depreciation ($/yr)	Return on investment ($/yr)	Cost flux (¢/h)
1 Pump	1071	2250	41.5
2 Exchanger	1840	6900	109.3
3 Flash tank	600	1800	30.0
4 Pump	929	1950	36.0
5 Mixer	3214	6750	124.6
6 Exchanger	1286	2750	50.5
Total			392.0

Thus, the combined costs to own and operate the process can be shown for each segment. The trade-offs for capital and available energy cost are made more obvious. For example, segment 6 has the highest energy cost and a relatively low capital cost. Improving the thermodynamic efficiency of this segment by reducing the LMTD would be a rather obvious place to explore for process improvement.

In the heater of segment 2 there are lower incentives to increase the heat transfer area, because the capital component is already high and the dissipation is lower. That is not to say that the system cannot be improved economically, however. Reducing the pressure in this heat exchanger would have two beneficial effects: it would reduce the capital cost of the heater and the pump and it might make possible the use of low-pressure steam, a utility that is priced somewhat lower on an available energy basis than high-pressure steam. We can infer that fouling is a problem in this heater when vaporization occurs because of the extent to which the designer has gone to prevent it. A different type of heater or some form of mechanical renewal of heat transfer surface (e.g., scraped surface exchangers) may be needed to achieve vaporization without excessive fouling. The incentives to carry out this development can be derived from the diagram.

The best approach to improving the operations of the flash tank would be to eliminate it in conjunction with an improved heater design. The ideal available energy change associated with separating the water from the solution must still be provided, but this is generally small compared to the actual work lost in separation processes.

The intent here is to demonstrate how the thermoeconomic data are used to focus on the process segments with the most potential for improvement. Compared to simple thermodynamic analysis, the combination of variables leads the designer more quickly to the key issues. Note also that the capital data are very sensitive to depreciation rates and required return criteria. Thus, different directions could well be suggested if these assumptions were changed.

IV. Systematic Optimization

Both of the approaches described provide data on which a process engineer can base some creative thinking. Many technical managers desire a system that points the direction to improvements mathematically rather than one that relies on the intuition of sometimes inexperienced engineers. M. Tribus and Y. M. El-Sayed of the Massachusetts Institute of Technology,

R. Evans and co-workers at the Georgia Institute of Technology, and other researchers of optimization systems are developing approaches that would do just this. These systems involve some sophisticated mathematics and are necessarily complex. They build on the thermoeconomic accounting concept and optimization theory to accomplish the objective.

A. Describing the Plant System

The first step is to develop available energy (exergy, essergy) balance equations and the direct prices for available energy described earlier. The "dead state" can be either ambient or another "valueless" condition.

Thereafter, "costing" equations are developed. These equations transform the cost of ownership of process equipment based on geometric variables that describe the equipment to costs based on process variables. These variables include those required for thermodynamic analysis. For example, a heat exchanger *cost* equation would depend on *geometric* variables such as tube length, number of tubes, material type and thickness, number and size of bolts, number of baffles, etc. A heat exchanger *costing* equation would depend on such *process* variables as the inlet and outlet temperatures, pressures, flows, and compositions for both the hot and cold streams. The cost equations will have more variables than the costing equations. The residual variables must be adjusted to maintain the costing equations at a minimum.

To avoid undue complexity, the engineer would include only the most important geometric variables and the most important process yield and energy variables. Selection of the required process variables is crucial. A different costing equation for each piece of equipment will result, depending upon which process variables are fixed and which can vary. A major advantage of such a set of equations would be to make the impact of the process variables on the cost of equipment more clearly apparent to the process engineer.

As in earlier approaches, the entire process under consideration is divided into individual segments; the cost of ownership of the equipment in each segment is determined, the cost of the available energy used in each segment is computed, and these two costs are compared. Segments with high ratios of available energy costs to ownership costs are made focal points for modifications. A modification in one segment may influence another segment either by causing a modification and consequent capital cost change in the second segment or by influencing the cost of available energy the second segment. Modifications thus result in a requirement to repeat portions of the analysis.

B. Thermoeconomic Optimization

As in other optimization problems, the approach is to find the minimum for the costing equations subject to the constraints imposed by the problem. This can get quite complex.

The costing equation can, in theory, be defined for each piece of equipment in the plant. The independent variables defining the costing equation are the process variables that control the physical design of that piece of equipment (temperatures, pressures, flow rates, compositions). *Only the variables that are required for later analysis are allowed to remain in the costing equation for each equipment item.* This costing equation is then minimized with respect to all other variables. When the resulting costing equations for all the equipment in the plant are added, a new equation results that gives the cost of ownership of the total plant in terms of a selected set of process variables.

All energy flows are defined in terms of available energy. The dissipation of available energy depends on the values of the process variables. The energy flows to each zone (or major piece of equipment) are priced according to the cost of available energy entering the system, as discussed earlier. This results in a set of available energy cost equations to go with the equipment costing equations.

Another set of equations, called the "constraint" equations is then developed. These equations define relationships among the process variables that must be satisfied. They are based on heat balance, material balance, and thermodynamic or engineering relationships (such as those describing the dependence of a stream's temperature rise on the heat input, or the representation of pressure drop versus velocity, etc.). Constraint equations must also be developed for external variables such as required product rates, compositions, and temperatures.

C. A Mathematical Approach

The cost of ownership and cost of available energy equations are added together to form a new equation (called the objective function) that is to be minimized subject to the limitations imposed by the constraint equations. This is a conventional (but magnificent) nonlinear optimization problem.

If a system is described by a set of variables,

$$\{V_s\}, \qquad s = 1, 2, \ldots, m,$$

and a set of constraint equations as described previously,

$$\phi_j(V_s) = 0, \qquad j = 1, 2, \ldots, n, \tag{11-20}$$

with $m > n$, then the approach to minimizing the cost of owning and operating the plant (the objective function) would be to assign optimum values to $m - n$ variables and solve the set of equations

$$\{\phi_j = 0\} \tag{11-21}$$

for the others. In Lagrange's original approach, all variables were treated the same way. Tribus and El-Sayed[11] proposed portioning the variables into two classes:

(1) "decision variables," (Y_k), $k = 1, 2, \ldots, (m - n)$, which give the designer a feel for how the system behaves;
(2) "state variables," (x_i), $i = 1, 2, \ldots, n$.

There are no set rules about which variables are classified in which category. It has been found useful to have the parameters familiar to the designer, such as efficiency, pressure ratio, etc., classified as decision variables in each segment. Thus, in any segment the state variables might be pressure, temperature, and work input, whereas the decision variables might be compression ratio and cycle or exchanger efficiency. Switching variables from one category to the other is generally possible *if* the degrees of freedom in the system are not reduced.

To understand the concepts described next, an understanding of the minimization process is needed first. Minimization occurs when all of the process variables are set such that a change in any of the process variables *increases* the value of the objective function (cost). Mathematically, this occurs when the differential of the objective function is zero. This is called a *necessary* condition for minimization. However, because the differential of the objective function is also zero at a maximum or saddle point, the zero value is not a mathematically *sufficient* condition for minimization. Analytically (through equations) proving a minimum is difficult in a multivariable problem. But numerically it is relatively easy just to change a variable and find out whether the objective function increases or decreases. If, when each variable is changed slightly one by one, the objective function always increases, the function is at a minimum. However, if it decreases when one or more of the variables is changed, the point is a saddle point and a lower cost exists. If it decreases for *all* variables, then the function is at a maximum.

The approach taken by Tribus and El Sayed involves the use of Lagrange multipliers. This is a complex technique wherein each constraint equation is multiplied by a factor (the Lagrange multiplier) and the result is added to the objective function to form a new function. This new function is then optimized by searching for a set of values that makes its derivative equal to zero. The derivative of the new function at any point away from the optimum is also taken as an indicator of the direction of change in the objective function resulting from a change in any variable, and thus as an indicator of whether that change reduces or increases costs.

Practical applications of this approach have yet to be developed in sufficient detail to allow the practicing engineer to determine whether the mathematical complexity of the optimization approach is justified by the design improvements it identifies. Meanwhile, other investigators are working on what may be less complicated routes for optimizing very complex nonlinear systems of equations. Clearly, this is very much an area for research as opposed to a proven approach for industrial application.

Thermoeconomics Summary

The approach of combining thermodynamic and economic data into an appropriate single criterion for identifying and prioritizing the sources of profit loss in a process has considerable potential. For relatively simple (or simplified) cases, it is possible to carry out this analysis by hand, with not much more effort than is necessary for an available energy analysis.

Mathematical techniques for optimizing complex thermoeconomic systems are still in the early stages of development and need to be proven in practical cases before being adopted by industrial process engineers.

Notes to Chapter 11

1. D. Boland and B. Linnhoff, The preliminary design of networks for heat exchange by systematic methods. *The Chemical Engineer,* Issue No. 343, 222–228 (1979).

2. J. W. Ponton and R. A. Donaldson, A fast method for the synthesis of heat exchange networks. *Chemical Engineering Science* **29**, 2375 (1974).

3. T. Lamaeda, T. Harada, and K. Shiroko, A thermodynamic approach to the synthesis of heat integration systems in chemical processes. *Computers & Chemical Engineering* **3**, 273–282 (1979).

4. T. Lamaeda, J. Itoh, and K. Shiroko, Heat exchange system analysis. *Chemical Engineering Progress* **74**, 70 (1978).

5. E. C. Hohmann, "Optimum Networks for Heat Exchangers." Ph.D. Thesis, Department of Chemical Engineering, University of Southern California, Los Angeles, California.

6. B. Linnhoff and J. R. Flower, Evolutionary generation of networks with various criteria of optimality. *AIChE American Institute of Chemical Engineers Journal* **24**, (4) 642–654 (1978).

7. T. Umeda, T. Harada, and K. Shiroko, A thermodynamic approach to the synthesis of heat integration systems in chemical processes. *Computers & Chemical Engineering* **3**, 273–282 (1979).

8. D. W. Townsend and B. Linnhoff, Designing total energy systems by systematic methods. *The Chemical Engineer,* Issue No. 378, 91–97 (1982).

9. The author shall refer to work carried out by M. Tribus of the Massachusetts Institute of

Technology and R. Gaggioli, formerly of Marquette University, and their co-workers. A number of other investigators are also contributing to this field and are not meant to be slighted. The limitations of time and space in a general text such as this force the author to stick to concepts.

10. R. Gaggioli and W. Wepfer, "Available Energy Accounting—A Cogeneration Case Study." Paper presented at the 85th National Meeting of the American Institute of Chemical Engineers, Philadelphia, Pennsylvania, June 1978.

11. M. Tribus and Y. M. El-Sayed, "Thermoeconomic Analysis of an Industrial Process." MIT Center for Advanced Engineering Study, Massachusetts Institute of Technology, Cambridge, Massachusetts, December, 1980.

12

Guidelines and Recommendations for Improving Process Operations

Introduction

In a pioneering paper on the calculation of the second law efficiency of chemical processes, Denbigh[1] made the following recommendations for increasing process efficiency.

(1) All heat transfer should take place at the least possible temperature difference.

(2) Pressure drops and other forms of friction should be kept at a minimum.

(3) Chemical reactions should be carried out under "resisted" conditions so that they will yield useful work.

These recommendations are very broad. Most process engineers already practice these guidelines within the limits of cost–benefit economics. This chapter provides more specific ideas for improving individual process operations and equipment without the need to carry out detailed thermodynamic analyses. In short, an attempt is made to distill some of the insight to be gained from analysis into some useful rules of thumb. However, it should be noted that the economics of energy efficiency improvement projects is a strong function of both the plant environment and time. The recommendations to maximize or minimize specific process conditions must be tempered with the economic limits imposed by the plant location and time period.

The guidelines will be broken down into several categories. Some of these will apply to individual unit operations or processes, and some will be aimed more at system interactions. Although every attempt will be made to be as specific as possible, there will be no escaping the overtone of motherhood. It is hoped that readers will be able to apply some of the general insight summarized in this chapter to their own process improvement efforts.

There is one overriding guideline that must be practiced if all of your efforts are to have practical value: *select programs and projects that will be implemented.* Too many sophisticated and comprehensive energy studies are either incomplete or gathering dust on a shelf because there was no commitment for implementation developed along with the technical details. Thus, it is much more beneficial to complete a sloppy energy audit and analysis and implement one energy-saving idea than it is to devise a brilliant and precise energy plan that goes nowhere. In the words of some philosopher somewhere, "if it's worth doing, it's worth doing badly—for a start."

I. Chemical Reactions

Often chemical reactions represent a major source of irreversibility in the process industry. Sometimes it is most practical to accept these losses as inherent because of the magnitude of process development needed to invent an alternative. However, even inherent reaction losses can be minimized within economic and operability constraints if the following guidelines are applied with judgment.

1. Minimize the Use of Diluents in Both Exothermic and Endothermic Reactions

The most common exothermic reaction is the combustion of fuel in furnaces and boilers. The presence of excess air, although essential for complete combustion, reduces the flame temperature. By reducing the ratio of air to fuel in combustion, one can increase the flame temperature and reduce the amount of relatively hot flue gas leaving the system. The availability of reaction heat at higher temperature and reduced heat losses from the flue gas enhance the energy efficiency of the furnace and boiler systems.

Another means of enhancing the quality and recovery of combustion heat is to use oxygen or oxygen-enriched air. However, complete air separation is very expensive. Partially-enriched air may be appropriate in some cases. This is particularly true when the process can utilize the higher flame temperature in conjunction with the reduced flue-gas losses. Diluents also affect available energy utilization and recovery in other exothermic reactions. The reactor example and phthalic anhydride problem in Chapter 9 show this.

In the case of endothermic reactions, the diluents increase the demand for high-quality heat. The quality of the heat supplied is significantly degraded in the heat recovery steps downstream from the reactor owing to the temperature differences involved in heat transfer. Thus, it is more efficient to keep the diluents at the minimum possible level as dictated by the reaction constraints.

The reduction of diluent in a reactor not only reduces the energy required but also increases the reactor capacity.

2. Maximize Product Yields in the Reactors to Minimize or Eliminate Separations, Recycle Streams, and Raw Material Waste

The benefit from increasing the product yields is that the energy requirements in downstream separations are significantly reduced. In addition, the equipment downstream from the reactor can be much smaller.

It is worthwhile to note here that chemical research programs designed to optimize reaction conditions and develop improved catalysts will have a significant effect on the energy efficiency of the manufacturing operations. For example, a better approach to minimizing energy requirements in difficult separations in ethylene plants might be to develop a means for increasing the ethylene content in the effluent gas from the cracking reactors.

3. Carry Out Exothermic Reactions at the Highest Possible Temperature and Endothermic Reactions at the Lowest Possible Temperature

This guideline helps to increase the recovery of higher-level heat from exothermic reactions and minimize the quality of energy required to drive endothermic reactions. Better catalysts and/or more economic reactor designs are generally required to make progress in this area. Please note, however, that previously optimum reactor conditions may well need updating at current and future hydrocarbon prices.

4. Consider Combining Exothermic and Endothermic Reactions to Minimize Losses

When the opportunity presents itself, combining reactions in a single interactive reactor has the potential to reduce losses. Only a single loss is incurred instead of two separate and additive ones. The simplest example would be to cool an exothermic reaction with feed that needs preheating or makeup boiler feed water. Such arrangements may not lead to the best utilization of available energy, because little flexibility is possible in matching sources and sinks. Nevertheless, for cases in which a good match of available energy needs and availabilities can be arranged, losses can be significantly reduced.

II. Separations

Various separation process losses are also often considered inherent. Efforts are made to reduce losses in heat exchange and other auxiliary processes, but the losses in the prime separation step itself are often accepted without question. As we have shown, the unavoidable loss in separation is often a very small part of the actual lost work in the step. The following guidelines can reduce losses to a point closer to the truly inherent.

1. Minimize the Mixing of Streams with Different Compositions, Temperatures, or Pressures

The demethanizer feed temperature example demonstrated this rule effectively, but it has two corollaries.

(1)   Inject feed at the point in the process when its composition matches the process composition.

(2)   Do not overpurify and then mix back. Separation energy requirements go up rapidly as purity approaches 100%.

2. Carry Out Separations at Conditions that Maximize the Separation Potential

For example, many fractionation processes have increased relative volatilities at lower pressures. As a result, lower reflux ratios and fewer trays are needed. For separations of hydrocarbon molecules lighter than hexane, both lower energy and lower capital costs are possible at lower pressures.

3. Incorporate Interreboilers and Condensers in Distillation Towers

This guideline helps one to use lower-level heat in inter-reboilers and to recover higher-level heat from intercondensers. In distillation towers operating at subambient temperatures, intercondensers allow the use of a less valuable refrigeration level and interreboilers allow the recovery of more valuable refrigeration.

*4. Maximize Heat Reuse in Distillation Towers by
   Using Heat Integration, Multiple Effect Systems, or
   Heat Pumps*

Heat reuse, e.g., using a tower condenser as a reboiler for another tower or as a steam generator, reduces the need for prime heat sources to reboil distillation towers.

*5. Maximize the Number of Stages in Multistage
   Separations such as Distillation, Extraction, and
   Absorption*

Use of more stages reduces the reflux ratio and thus reduces energy requirements.

*6. Carry Out Difficult Separations in the Absence of
   Non-key Components*

By eliminating the non-key components from a distillation operation, one reduces the temperature difference across the distillation tower. Thus, the degradation of heat energy (availability loss) is reduced.

III. Heat Transfer

Much has already been discussed about the sources of lost work in heat transfer. We shall briefly summarize some guidelines.

*1. Minimize the LMTD in Heat Exchangers,
   Furnaces, and Boilers*

Lower temperature differences in heaters or coolers mean recovering waste energy at the highest available energy content and using the lowest quality energy input. In refrigeration systems, following this rule leads to cooling with the warmest (lowest available energy level) refrigerant and recovering refrigerant at the coldest level. The following recommendations can be derived from this guideline.

(1) Use the lowest possible level steam in process heaters.
(2) Reduce or eliminate combustion whenever possible, e.g., replace direct-fired reboilers in distillation towers with steam heaters.
(3) Use multilevel refrigeration cycles in low-temperature cooling. Consider the use of multicomponent refrigerants to minimize LMTDs in low-temperature exchangers.

(4)   Develop practical topping cycles for furnaces and boilers to eliminate large temperature differences and recover work.

2. Minimize the Use of Direct Heat Transfer Operations such as Quench Operations

Quench operations lower the quality of the heat source. Thus, the potential for the recovery of higher-level heat is lost. Direct-contact quench
coolers are desirable when the primary objective is to cool the process
streams to the lowest possible temperature, because they achieve lower
approach temperatures.

3. Set Approach Temperatures as Functions of Temperature Levels

Economic analysis has shown that refrigerated systems should be designed with much closer approaches than those at near-ambient temperatures. The impact of the steam power system investment is significant in
these cases. Considerable room exists to optimize even cooling water exchangers, taking into account the cost of the cooling tower systems. Old
rules of thumb are generally no longer optimum.

IV. Process Machinery

Much of what has been said about separations can be translated to
machinery. A significant difference is that current machinery efficiencies are
much higher than those for separation processes. Therefore, less improvement per unit of consumption is possible, but consumptions are generally
very much larger. The net result is that the overall economic impact can be
quite large.

1. Maximize the Number of Stages in the Compression

Thermodynamic calculations show that the horsepower required in a
compressor goes down as the number of stages in the compression is
increased, up to some limit.

2. Use the Most Efficient Pumps, Compressors, Turbines, Motors, etc.

It is obvious that more efficient process machinery would help reduce
both the use of more valuable mechanical or electrical energy and the
investment needed to provide it.

3. Use Expanders Instead of Throttle Valves

Expanders allow the recovery of valuable mechanical work. Also, they provide additional cooling of process streams. The use of an expander on demethanizer overhead streams in ethylene plants is common.

4. Minimize Pressure Drops in the Processing Units and Control Systems

Valuable mechanical or electrical work consumed in pumps and compressors can be saved by reducing pressure drops. Pressure-drop reduction usually also results in increased plant capacity. In some cases reducing pressure drops can lower reaction pressure and increase yield (as in ethylene cracking furnaces) or increase compressor suction pressure to decrease compression requirements. However, the impact on utility systems must be accounted for.

5. Select Drivers on the Basis of Integrated System Analysis

Depending on the relative amounts of process heat and power required in a plant, different types of drivers make greater or lesser overall plant fuel efficiency possible. An overall steam power system study is needed to select – optimize the combination of operating drivers as conditions change.

V. System Interactions and Economics

Considerable time has been spent in this book demonstrating that the entire plant system must be taken into account in analyzing fuel (available energy) consumption. Any of the preceding guidelines are subject to limitations imposed by interactions with other systems. Some points to be aware of in this area follow.

1. Check that Each Process Step Is Necessary before Trying to Improve It

Eliminating a step eliminates both the ideal work of the step and the associated irreversibilities. It represents a 100% saving, provided that the impact on other parts of the process does not consume the benefit. Elimination of H_3PO_4 evaporation in the hemihydrate process of Chapter 10 is an example.

2. Upgrade the Impact of an Available Energy Input

The most attractive improvements often involve the use of less available energy to have the impact of more. For example, combustion air might be heated to 350°F by waste heat at 400°F. The net impact is to get back out the same number of Btu's of fuel from the furnace or boiler. In effect, an equal quantity of 400°F heat has been traded for fuel that has a much higher available energy and large inherent losses in use.

The same reasoning works for refrigeration systems. If process conditions are changed so that the same amount of −50°F refrigeration is used to replace −100°F refrigeration, a net compressor horsepower saving will result. Depending on the steam power system, this saving may be cascaded back through the compressor driver and utility system to obtain a fuel impact that is significantly larger than the original saving.

3. Minimize the Downgrading of Available Energy

Several commonly accepted energy recovery techniques actually do this in an effort to recover more Btu's (increase first law efficiency). For example, waste heat boilers are positioned on furnace stacks and internal combustion engine (gas turbine and diesel engine) exhausts. Generally, steam at 600 psig or less is generated from flue gas as hot as 900°F. The net result is to produce an energy product at 500°F from a source at 900°F. This represents an appreciable loss of available energy. Regenerators might be a better substitute.

Control systems inherently degrade available energy for another desired purpose: steam is let down across valves for pressure control. Throttle valves control pump flow rates. Flow meters create a pressure drop and smaller-sized pipes are used to save investment (sometimes mistakenly). Alternative methods to achieving many of these objectives (e.g., variable spread pumps) are possible, often at no increase in capital cost if the entire energy system is properly considered. Think about them.

4. Concentrate Dispersed Irreversibilities to Facilitate Efficient and Cost-Effective Recovery

Small, widespread irreversibilities make for economic and technical problems in recovery attempts. Many small, single-stage steam turbines may be difficult to upgrade in efficiency in an economic manner. Perhaps they should be replaced with one large turbogenerator and small motors. Heat recovery from a number of small sources (e.g., furnace stacks) may be difficult to justify. Perhaps they could be accumulated by ducting or creating a heat transfer loop into a single-recovery system designed to maximize the recovery of available energy. Conversely, perhaps one large, efficient furnace can be designed to serve several smaller process duties in a more efficient way.

5. Avoid Unnecessary Heat Exchange

This may be considered as a corollary to the first guideline in this section, but it has a broader impact than simply eliminating a step in a given process unit. Often some surge volume is provided between a series of in-line process units. For example, the gas oil or naphtha stream from a pipe still is usually cooled and stored before being fed to a steam cracker. Running the stream directly to the cracking furnaces eliminates the irreversibilities (and capital cost) associated with both cooling off the stream for safe storage and reheating it to feed the process. Clearly, control and safety problems could exist in this arrangement, but it is being done more frequently along with the development of appropriate contingency plans to decouple the two units when necessary. Similar possibilities may exist in your complex.

6. Consider the Entire Plant System When Evaluating Energy Efficiency Improvements

A great deal has been said on this subject. This is a final reminder that both the thermodynamic and economic (capital and operating cost) impacts of any proposed change must be evaluated all the way back to the point of primary available energy input to be sure that you have the right answer.

7. General Economic Priority List

The following *priorities* have been found for the profitability of industrial energy conservation activities.

(1) Tune operations and control better.
(2) Directly reuse waste energy.
(3) Use waste energy to save fuel or power.
(4) Use waste energy to raise steam.
(5) Use machinery to recover or upgrade waste heat to more valuable levels.

8. Be Sensitive to the Inroads Increased Efficiency Will Make on Operating Flexibility and Essential Safety

Additional controls may be required to negate the impact of more frequent excursions from optimum conditions. In any event, commitment from the operating group to any change or new system will be essential to success.

9. Look for Synergism with Neighboring Facilities and with Environmental Programs

As discussed in some detail, cogeneration opportunities may exist across the fence. Although mentioned only briefly, there are many common purposes between environmental control and energy efficiency improvement. Any step that saves fuel also reduces pollution. A company that recognizes this and participates in joint projects and programs will benefit.

VI. A Checklist of Energy Conservation Items

Interpretation of the preceding guidelines will vary from plant to plant. Both the type of process equipment and the type of management system will contribute to this variation. The following checklist for a petrochemical complex has been drawn up to illustrate one translation of the guidelines into a series of specific action steps. The list is not meant to be comprehensive. Some items involve investment opportunities, some only increased diligence by operations and maintenance personnel. The reader can develop a checklist for his own plant if that approach holds promise for improving profits.

1. Furnaces and Boilers

Is the furnace stack temperature at the minimum level permitted by the sulfur content of the fuel?

Have the furnaces been checked for air leaks?

Does the control house have a portable oxygen analyzer for balancing burner operation?

Are in-stack analyzers installed on furnaces larger than 10–20 MBtu/h and are others monitored regularly?

Is the fuel oil temperature high enough to ensure that the viscosity of the oil is below 120 Saybolt seconds (25 centistokes)?

Is the atomizing steam dry?

If the fuel gas is at the dew point, is there a knock-out drum at the furnace with steam tracing to the burners?

Are all soot blowers operating and maintained regularly?

Are all draft gauges operating?

Have furnaces and boilers been studied for the possibility of recovering additional heat by replacing bare convection tubes with finned or studded tubes, adding extra tube rows, or installing air preheaters or waste heat boilers and induced-draft fans?

Have the burners been checked for possible replacement with more efficient models?

Is high-pressure gas available to replace steam for the atomization of liquid fuels?

2. Heat Exchangers

Can heat integration between compatible units be improved?

Can additional exchanger shells for existing exchanger systems be justified?

Can low-fin tubes be installed in existing shells to improve recovery?

Have controlled-pitch fan hubs for air-fin coolers been considered?

Are preheat exchanger trains monitored to determine the optimum cleaning cycle?

Is the rate of cooling water flow controlled to minimize use?

Can extra shells be installed on reboilers so that exhaust steam can be used in place of high-pressure steam?

Is heat rejection to air or water limited to process temperatures below 250°F?

3. Insulation

Are hot flanges, manways, and pumps insulated with removable insulation to permit maintenance?

Have medium-temperature tanks been insulated?

Are the maintenance personnel familiar with current insulation economics?

Has a survey of all units been made to catalog insulation needs?

Have control valves in the refrigeration system been properly insulated and continuous derhyming been eliminated?

4. Fractionation

Have the towers been checked for correct tray design and feed tray location to increase the number of theoretical stages so that reflux can be reduced?

Have the fractionation systems been studied for the application of heat pumps, multiple effect schemes, and mid-reboilers?

Is stripping steam minimized in towers where direct steam injection is used?

Have the operating pressures been minimized to save reflux?

Are the reflux ratios adjusted so that the last increment of energy input is justified by the value of improved fractionation?

Have the sour water strippers been checked for the correct number of stripping and rectification trays?

5. Computer Control

Have the furnaces been considered with respect to controlling excess air?

Have the tower control loops been modified to keep reflux at a minimum level that is consistent with fuel prices and reduced throughputs?

Are the compressors controlled to minimize recycling and power requirements?

Are the heat exchange trains controlled to maximize energy savings?

6. Steam Power Systems

Is the venting of exhaust steam minimized by shutting down turbines and starting up motors?

Is the idling of standby turbines minimized?

Can tank-heating steam be replaced by exhaust steam, condensate, or other low-level heat source after the tanks are insulated?

Have the pumps been checked to see if smaller impellers can be applied to avoid recycling or excessive control valve pressure differences?

Is all of the condensate recovered?

Is the steam system kept in balance?

Are there any depressuring stations that can be replaced with steam turbines?

Is the boiler feed water preheated using process heat rather than by sparging exhaust steam or using cold feed water directly?

Is all of the exhaust steam recovered?

Can the steam drum internals be improved to reduce the blow-down rate?

Is flash steam from the blow-down recovered in the steam system?

Does the plant power factor need improvement?

Have all opportunities to generate power at less than 10,000 Btu/kW been explored?

Is steam generated at maximum pressure and let down through turbines to condensing pressure?

Have inefficient single-stage turbines been replaced with multistage machines?

Is the use of condensing steam turbines minimized?

Have turbine backpressures been minimized?

Are obsolete steam lines removed to avoid heat loss?

Can hot process or water streams be used to replace steam tracing?

Are steam depressuring station valves leaking when closed?

Has a routine steam trap maintenance program been implemented?

Is a steam ejector used to upgrade exhaust steam instead of throttling more higher-pressure steam when steam is needed at a pressure less than the line pressure?

Have the steam ejectors been checked to minimize steam use?

Are mechanical vacuum pumps a better choice than jets for some applications?

7. Flares and Offsites

Are individual lines to the flare metered or is an analysis of the gas made to determine the source and allow the correction of releases?

Are flare-gas compressors justified?

Can flare gas be recovered by a lean absorbent stream?

Are combustible waste products recovered in the incinerator or other furnaces (to reduce the auxiliary fuel)?

Is the sludge adequately dewatered to reduce disposal energy costs?

Has flare seal-drum pulsing been eliminated to steady the flow rate and reduce the steam requirement at the flare tip?

8. Miscellaneous

Has a process energy audit been carried out on all units?

Does the utility operator make routine checks for steam leaks around the plant?

Do operators monitor the energy consumption per unit of production?

Has an infrared temperature survey been conducted to determine the source of heat leaks and safety valve leaks?

Has adequate training been given to the Operations, Technical, and Maintenance Departments?

Has an in-plant "energy awareness" program been implemented for *all* employees?

Are buildings, hot water, and other low-level heating needs provided with exhaust steam, condensate, or other waste heat stream?

Have major process improvements needed to reduce energy consumption been brought to the attention of the technical staff?

Have all integration opportunities with neighbors been explored?

VII. Shortcomings of Guidelines

Any list of guidelines is meant to be a shortcut for more thorough analysis. As specific as they might sound, they are designed to apply to general situations. Their use may get some quick results, but they may cause you to overlook the optimum configuration specific to your own situation. There may also be inherent conflicts in any set of guidelines. These guidelines are based on fairly fundamental priniciples, but specific situations, such as locally cheap purchased power, may still introduce conflicts. There is no substitute for a thorough analysis of the interactions present in your own system.

A final caveat involves safety. Although it may sometimes be necessary to modify current safety practices to achieve a given efficiency improvement, no engineer or manager can afford to compromise the *essential safety* of his unit. Nothing in this book is intended to advocate sacrificing safety for efficiency. The safety aspects of any opportunity must be worked on harder than the thermodynamic ones.

Notes to Chapter 12

1. K. G. Denbigh, The second law efficiency of chemical processes. *Chemical Engineering Science* **6,** 1 (1956).

Index